THE ANARCHIST GEOGRAPHER:

AN INTRODUCTION TO THE LIFE OF PETER KROPOTKIN

BRIAN MORRIS

Genge Press

Genge Press, 45 Quay St., Minehead, Somerset, TA24 5UL, England
gengepress@aol.com; www.gengepress.co.uk

© Brian Morris 2007

The moral right of the author has been asserted

All rights reserved. No part of this work may be reproduced or transmitted in any form or by any means, electronic or mechanical, or duplication by any information storage or retrieval system without prior written permission from the publisher, except for the inclusion of brief quotations with attribution.

British Library Cataloguing in Publication Data
A catalogue record for this book is available from the British Library

ISBN: 0-9549043-3-8; 13-ISBN: 978-0-9549043-3-3

Typeset in Times New Roman and printed by Friday Print, Minehead, UK

Contents

	page
Introduction	v-vi
1. Boyhood and Youth	1-11
2. Travels in Siberia	12-22
3. Becoming an Anarchist	23-33
4. Russian Populism and the Chaikovsky Circle	33-45
5. The Jura Federation and Anarchist Communism	46-59
6. The Socialist Movement in England	60-73
7. The Coming Revolution	74-85
8. The Crisis	86-92
9. The End of Exile	93-105
Chronology of Important Events in the Life of Peter Kropotkin	106-7
Bibliography	108-112
Index	113-119

To the memory of NICOLAS WALTER

humanist, anarchist and scholar

Introduction

An important and talented geographer, an explorer in his early youth, Peter Kropotkin was one of the most seminal figures in the history of the anarchist movement. He has indeed been described as a unique combination of the prophet and the scientist. Although Kropotkin made many important contributions to science, particularly his theory of 'mutual aid', which emphasized the importance of co-operation and symbiosis in the evolutionary process, throughout his life he was a revolutionary socialist, devoting time and energy to the anarchist cause. By his exemplary life, and by generating 'a treasury of fertile ideas', as his friend Errico Malatesta put it, Kropotkin undoubtedly stirred the imagination of his generation. He was also a pioneer ecological thinker, and his *Fields, Factories and Workshops* was one of the great prophetic works of the nineteenth century.

Kropotkin has generally been ignored by academic scholars, who seem to prefer the obscurantist musings of such reactionary philosophers such as Heidegger, but Kropotkin's ideas continue to find resonance in many contemporary currents of thought – in the urban ecology of Lewis Mumford and Paul Goodman; in the bioregional vision of Kirkpatrick Sale; in the social ecology of Murray Bookchin; in the plea for intermediate technology and organic farming by the likes of E.F. Schumacher and Wendell Berry, and in Takis Fotopoulos's project of inclusive democracy, to name but a few. Even poststructuralist philosophers like Michel Foucault and Gilles Deleuze seem to have appropriated many of the ideas of Kropotkin (and other anarchists) – with very little acknowledgement – particularly Kropotkin's critique of the state, capitalism, representation and the vanguard party (Marxism).

A friend and close associate of William Morris, George Bernard Shaw, Edward Carpenter and the redoubtable Emma Goldman, who described Kropotkin as 'my great teacher', Kropotkin made enduring and substantial contributions to the development of physical geography and ecological thought, as well as to anarchist theory.

This short biography of Kropotkin has one purpose only: to keep alive the memory of an anarchist scholar and revolutionary socialist, and to introduce the reader to the life and times of Peter Kropotkin.

Brian Morris
June 10th 2007

For recent studies of Kropotkin's social ecology and revolutionary anarchism, see Graham Purchase: *Evolution and Revolution: an Introduction to the Life and Thought of Peter Kropotkin* (Jura Books, 1996) and my *Kropotkin: the Politics of Community* (Humanity Books, 2004).

One: Boyhood and Youth

It is one of the curious ironies of history that one of the fiercest opponents of all governments and of the centralised nation-state, should have been a direct descendant of the ruling dynasties that once governed his country (MK3)[1].For Pyotr (Peter) Alexeivich Kropotkin was born into the highest rank of the Russian landed aristocracy, his princely forebears having been among the early rulers of Russia. Prior to the consolidation of the Russian state in the early seventeenth century, under the Romanov dynasty, the Kropotkin family were rulers of the principalities of Smolensk and Kiev, or at least so the family tradition claimed. But by the nineteenth century the family was less concerned with government service than with managing their own large landed properties.

Kropotkin's father, Prince Aleksei Petrovich Kropotkin (1805-71) was a prosperous land-owner who owned three large estates, all of which were well-populated with serfs. Wealth in those days was measured in the number of 'souls' that a landed proprietor owned, for the peasants were the main producers, and were bonded as serfs to the land-owners; they could be sold like goods and chattels. There was an old Russian saying that 'the peasants belong to the lord, but the land belongs to the peasants' (WA 8). Kropotkin's father, on his three estates, owned nearly twelve hundred 'souls' – that is, adult males, as women and children were not counted.

Prince Aleksei Kropotkin was by all accounts a rich landowner, and he lived up to his reputation by frequently holding extravagant parties and entertaining large numbers of visiting

---

[1] Abbreviations within brackets refer to works listed in the Bibliography

nobility. Besides his country estates, Kropotkin's father also owned a large house, maintained by around fifty servants, in the old equerries quarter of Moscow. Situated behind the Kremlin, this quarter was favoured by the old Moscow nobility, and was far away from the noise and bustle of the city's commerce. It was here that Peter Kropotkin was born on the 9th December 1842, and it was here that he spent the first fifteen years of his life.

During the summer months the Kropotkin family moved to their country estate at Nikolskoye, a village about 160 miles from Moscow, in the province of Kaluga (*Kaluzh*). On the estate there were around seventy-five domestic servants: four coachmen attended the horses, five cooks prepared the meals, a dozen people waited upon the table, and the remaining servants took care of the general household (MR 28). During his boyhood Kropotkin greatly enjoyed the summers he spent on the country estate at Nikolskoye.

Kropotkin's father was a typical army officer during the reign of Nicholas I (1825-55), although he took part in few military campaigns. His only claim to military distinction was a medal awarded to him during the Turkish campaign of 1828, but this was through the bravery of his own personal servant. On his estate he dressed in military uniform and insisted on military decorum at all times; he even communicated with his servants through written 'orders', always signing himself 'colonel and commander'. He was, as Kropotkin describes him, a 'true military man', despising all forms of attire except that of the military uniform. Such men, Kropotkin remarked, 'were trained to perform almost superhuman tricks with their legs and rifles' (MR 9-10).

Although honest and less corrupt than many of his aristocratic contemporaries, Kropotkin's father was typical of his class - emotionally distant towards his own family (Kropotkin hardly saw him), and brutal and inhuman towards the servants and peasants on his estates. He often physically abused the peasants, had them flogged for any misdemeanours, and wielded almost total control over their lives. He arranged their marriages as he thought fit (usually against their own wishes), and sent any young man who met with his disapproval away on military service. Serfdom, as Kropotkin experienced it in his own home, was thus a form of tyranny (MR 49-56).

Kropotkin's attitude towards his mother was very different. In his memoirs she is described as a rather saintly figure. But Kropotkin hardly knew her, for she died of tuberculosis in April 1846 when aged only thirty-five. Kropotkin was then only three and a half years old. His mother Ekaterina was the youngest daughter of a Cossack army officer, General Nikolai Sulima, who came from the Ukraine. She had married Kropotkin's father with great pomp in 1831, although his father always bewailed the fact that she brought little to the marriage by way of dowry.

Kropotkin's mother was by all accounts a remarkable woman, with both artistic and intellectual talents. Many years after her death Kropotkin discovered some of her old papers in a stateroom of the country-house at Nikolskye. They included water-colour paintings and books filled with verses and poetry, including the historical ballads of the Russian poet Ryleev, one of the Decembrists whose work was banned by Nicholas I and who was hanged in 1826 (MR 12). Ekaterina had three children besides Kropotkin: Nicholas, the eldest, was born in 1834, Helene the following year, and then, after a lapse of six years, Alexander was born in 1841. Kropotkin was thus the youngest child. Inevitably, all three boys were encouraged by their father to enter military service. In 1853, at the outbreak of the Crimean War, Kropotkin's brother Nicholas joined the army, and Kropotkin never saw him again (MR 63).

In 1848 Kropotkin's father remarried. His new bride was the daughter of an admiral in the Black Sea fleet, Elizabeth

(*Yelizaveta*) Korandino. From the outset her presence in the family household generated hostility and tension, particularly in relation to the two youngest boys, Alexander and Peter. This was not only because she was a rather domineering and bad-tempered person, but also because she had a rather vindictive streak. She not only removed from the house everything that might remind the children of their mother - her portraits, paintings and embroideries - but forbade them to have any contact with their mother's relatives. Not until much later in life did Kropotkin meet up with his maternal cousins.

But on the whole, Kropotkin had a fairly happy childhood, and in his memoirs three things are recorded as being of particular significance to him. These were his French tutor, his contact with the peasants on the estate, and the close relationship he felt towards the natural world.

As was common amongst the nobility of his day, Kropotkin was educated by private tutors, and a great emphasis was put on acquiring French culture. One of his tutors was Monsieur Poulain. He was one of the 'debris' of Napoleon's Grand Army who had stayed on in Russia after 1812. Besides teaching them the French language and etiquette, Poulain also instructed Kropotkin and his brother Alexander in grammar, history, mathematics and geography. His methods were rough and ready, and he liberally used the birch-rod, until Kropotkin's sister Helene intervened. But outside the classroom and during afternoon walks in the woods, his French tutor would relax his approach, and regale Peter with stories of French history and the French revolution. These accounts greatly stirred Kropotkin's imagination. He long afterwards recalled what his tutor had said about Count Mirabeau, who had renounced his title and 'to show his contempt for aristocratic pretensions, opened a shop decorated with a signboard which bore the inscription "Mirabeau tailor"'. Shortly afterwards, and by his twelfth year, Kropotkin had dropped the title 'Prince' and began signing himself P. Kropotkin (MR 47).

Perhaps because as a boy Kropotkin was somewhat estranged from both his father and stepmother, he came into close contact with the peasants, particularly with the domestic servants. He spent much time in their company, often joining in their social activities, and it was peasant women who nursed him as a child. In many ways he came to identify with the peasants - as against his own aristocratic class. In his memoirs he wrote:

> Few know what treasuries of goodness can be found in the hearts of Russian Peasants, even after centuries of the most cruel oppression, which might well have embittered them (MR 46).

Kropotkin's other attachment was to the natural world, and he wrote to his brother Alexander that 'nature was always my best friend' (Lebedev 1932: 1/66, MK 271). The countryside around the Nikolskoye estate made an indelible impression upon Kropotkin, and the beautiful pine forests near the estate remained in his memory: 'Some of the happiest reminiscences of my childhood'. He wrote:

> Immense red pines, centuries old, rose on every side, and not a sound reached the ear except the voices of the lofty trees ... Noiselessly a squirrel ran up a tree, and the underwood was as full of mysteries as were the trees. In that forest my first love of nature and my first dim perception of its incessant life were born (MR 43-44).

Having come to the attention of Tsar Nicholas I while attending a fancy dress ball organised by the Moscow nobility (when he was only eight), Kropotkin was selected as a candidate for the prestigious Corps of Pages. In August 1857, when he was almost fifteen, Kropotkin joined this privileged corps. Only one hundred and fifty boys, mostly children of the nobility belonging to the court, were educated in the corps. It was a unique organisation, combining the character of an elite military school, endowed with special rights, with that of a court institution attached to the imperial household. Kropotkin's father was absolutely delighted at his son's selection, as training at the school inevitably opened up many opportunities, offering 'every facility for making a brilliant career in the service of the state'. Kropotkin, however considered it 'a misfortune' that he had to enter the school (MR 70-71).

Kropotkin himself has graphically described in his memoirs the 'inner life' of the Corps of Pages: the severe discipline; the despotic nature of many of the instructors, and the harassment and sexual abuse meted out by the older boys, the select *Pages de Chambre* who were personally attached to the emperor. But happily for Kropotkin, Russia was then undergoing a spirit of reform. Alexander II had become Tsar in 1855, and there was a feeling that the 'terrible nightmare' of Nicholas I's reign was over (MR 73). So although Kropotkin resented the boredom, the senseless discipline, and the despotic ethos of the Corps, he began to relish the educational aspects of the military school, as many of the tutors were university professors. His intellectual interests had already been aroused while still at home by his Russian teacher Nikolai Smirnov. This tutor had introduced him to Russian literature (much of it forbidden), especially the works of Gogol and Pushkin. Kropotkin imbibed Smirnov's passionate devotion to Russian literature, and also began to write his own journal - a monthly review he titled *Vremenmk* (chronicle).

Now, at the Corps of Pages, Kropoktin's intellectual interests expanded and deepened through his encounters with many sympathetic tutors. A Professor Klasovsky, himself a classical scholar, not only taught Kropotkin Russian grammar, but also introduced him to philosophy and literature, as well as to folklore, ranging from Homer to the Mahabharata. Another professor, Herr Becker, who was librarian at the imperial (national) library, taught him German, and introduced him to Goethe's *Faust*, which gave Kropotkin 'unfathomable' joy. Through his sister Helene, who was married and living in St Petersburg, he gained access to the excellent library that her husband possessed. On Saturday visits, Kropotkin would immerse himself in the writings of the French philosophers of the eighteenth century, especially Voltaire, as well as the pantheism of the Stoics.

Professor Klasovsky had an immense influence on Kropotkin. 'Every school in the world', he wrote, 'ought to have such a teacher'(MR 88). Kropotkin followed the broad philosophical and humane teachings of his teacher, but he also developed his own interests in the natural sciences. Mathematics, physics and astronomy were, he noted, his chief studies.

Another important feature of Kropotkin's early education was the summer camps of the Corps at Peterhof, for although he hated the discipline and military drill at these camps, they gave him ample opportunity to explore the forests and countryside and to undertake mapping and survey-work, which to Kropotkin was always a 'deep source of enjoyment' (MR 124).

During his five years at military school with the Corps of Pages (1857-1862), Kropotkin corresponded extensively with his brother Alexander, who was at that time with the Cadet Corps in Moscow. The relationship between the two brothers was always close and intimate, though they differed greatly in personality and interests. Alexander was a poet, and had pronounced philosophical inclinations. He was introspective and interested in abstract ideas, and read both Kant and Hegel with enthusiasm. But Kropotkin had a much more practical and scientific bent, and though greatly interested in philosophy, poetry and the arts - he became an accomplished painter and musician - he never took to German philosophy. He much preferred the rationalist philosophers of the French enlightenment. Kropotkin found Kant difficult to understand, and like Herzen, never had much sympathy for Hegelian metaphysics (WA 33-34).

Kropotkin also differed from his brother Alexander in his political outlook, which was already moving in a libertarian direction, although he would not become an anarchist until 1872, when he was thirty. Alexander had become a close personal friend of Peter Lavrov (1823-1900), a radical socialist, who, as an exile in Geneva, would in 1873 edit the important radical newspaper *Forward!*. Lavrov was a great admirer of Marx, and like him was a social democrat as well as being a leading 'populist' intellectual. Alexander's political views were thus slightly more conservative than those of Kropotkin's, but still, in the eyes of the government, highly subversive (MK 34).

During his time with the Corps of Pages Kropotkin spent many hours undertaking military drill and manoeuvres, and learning the basics of military science, all of which had the aim of making him into a disciplined military officer. Like his brother Alexander, however, Kropotkin had long ago come to the conclusion that such a career was worthless and to be despised. He described it as an 'absurdity'. But it was also during his teenage years, prompted by this same brother's admonition that 'one must have a set purpose in life', that Kropotkin developed his own philosophical outlook. It was one that was neither religious nor a crude form of mechanistic materialism, but rather one, drawing on the work of the Stoics, Goethe, Humboldt and Darwin, all of whom he admired, that can be described as 'ecological': a social ecology. Kropotkin graphically outlines the ecological vision or ontology which was then emerging in his memoirs.

> The infinite immensity of the universe, the greatness of nature, its poetry, its ever-throbbing life, impressed me more and more; and that never-ceasing life and its harmonies gave me the ecstasy of admiration which the young soul thirsts for, while my favourite poets supplied me with an expression of that awakening love of mankind and faith in its progress (MR 97)

Poetry and science, music and chemistry, for Kropotkin went hand-in-hand. He continues:

> The never-ceasing life of the universe, which I conceive as life and evolution, became for me an inexhaustible source of higher poetical thought, and gradually the sense of man's oneness with nature, both animate and inanimate - became the philosophy of my life (MR 117).

Although Kropotkin was always a great advocate of modern science, such pantheistic sentiments, echoing those of Goethe, make it quite misleading to interpret Kropotkin as if he were a Cartesian duellist, or a 'mechanistic' philosopher of the seventeenth century - still less a crude positivist. Kropotkin's 'integrative' philosophy affirms not a form of positivism, but a 'poetry of nature'. It was thus quite understandable that he should be critical of the scholastic education of his own day, where the school was modelled on that of the medieval monastery (MR 125), and came to stress the importance of teaching the natural sciences: physics, astronomy, zoology and botany. What was needed, he felt, was an integrative natural philosophy along the lines suggested by Alexander Humboldt in his *Cosmos*:

> The philosophy and poetry of nature, the methods of all the exact sciences, and the inspired conception of the life of nature must (be a) part of education. (MR 89).

While Kropotkin was at military school, Russia was undergoing a great social transformation. Ever since the revolutions of 1848 in Western Europe, there had been sporadic outbreaks of student unrest and peasant revolts, while revolutionary ideas had been spreading throughout Russia in the form of novels or political tracts. As all forms of radical thought were severely censored by the authorities, any criticism of Tsarist autocracy had to be expressed in disguised form, or smuggled secretly into the country. The writings of the great exile, Alexander Herzen (1812-70) were particularly important. His loud and passionate critiques of the Russian autocracy were expressed in his books and in his review, *The Polar Star*, which was published from London. These writings were widely but secretly circulated throughout Russia, especially in St Petersburg, and made a great impact.

So throughout the early part of the reign of Tsar Alexander II (1855-1881) there were great debates on the need for a constitution and for the abolition of serfdom. As Kropotkin's biographers put it: 'The discontent among the peasants themselves, the guilty feelings of the more liberal nobility, and the spirit of radicalism that had been released by the death of Nicholas I', all combined to bring the issue of serfdom to the forefront of politics (WA 38). Finally, in February 1861, following the words of Herzen that it was better that the revolution 'should come from above' rather than be initiated by the peasants themselves, Alexander II proclaimed the emancipation of the serfs.

Initially Kropotkin had admired Alexander II, and, as the liberator of the serfs, the Tsar was at first greeted with enthusiasm throughout Russia. But it soon became evident to Kropotkin that these reforms were something of a sham, for the high redemption tax made economic freedom illusory. Many peasants, though they had personal freedom, had little else besides and they were often worse off than before their emancipation. The peasants on one of his father's estates (Tambov), he noted, were now paying rent to his father which 'represented twice as much as he used to get from that land by cultivating it with servile labour' (MR 138). Kropotkin was further disillusioned by state suppression of the voluntary schools that had spontaneously arisen in St Petersburg immediately after the abolition of serfdom, and by the savage response both to the student demonstration at the universities of Moscow, Kazan and St Petersburg, and to the insurrections in Poland. Alexander II, he noted, was to 'drown the Polish insurrection [1862-63] in blood'.

Having met the emperor personally on several occasions, Kropotkin came to the conclusion that Alexander II was a 'reactionary character' and a 'despot' (MR 148-51). He had become ever more conscious of the brutality of the Tsar's regime. By the time Kropotkin had completed his military training, St Petersburg had assumed, with the 'triumph of reaction', a rather gloomy atmosphere. He lost no time in getting away from the capital (MR 167).

Two: Travels in Siberia

Kropotkin graduated from the Corps of Pages in June 1862, having been nominated in his final year Sergeant of the Corps. One of the privileges of belonging to this elite military school, was that its graduates had a choice of regiments. Much to the consternation of his father and the astonishment of all his comrades, Kropotkin decided to join the newly formed mounted Cossack regiment of the Amur, in remote Eastern Siberia. That someone who was a sergeant of the country's most prestigious military school, and a personal page of the emperor, should not want to serve in one of the elite regiments stationed in St Petersburg, completely baffled everyone. His father telegraphed the director of military schools, the Tsar's brother the Grand Duke Michael, forbidding Kropotkin's move to Siberia. But fortunately, because Kropotkin had played a commendable role in controlling the St Petersburg fire of May 1862, the Grand Duke had shown an interest in Kropotkin's plans and had given him a letter of introduction to the governor-general of the Amur. This seems to have quelled his father's disapproval.

After visiting his family in Moscow, Kropotkin left for Siberia in August 1862. The Amur region had recently been annexed by Russia, largely through the efforts of General Nicholas Muraviev. What clearly motivated Kropotkin to visit the area was his reading of the works of such pioneer geographers as Humboldt and Ritter. He saw Siberia not only as a place where 'great reforms' could be achieved, in contrast to Russia itself, but also where he could undertake geographical explorations. During the next five years (1862-67), it is estimated that Kropotkin travelled around fifty thousand miles, by cart, steamer, boat or on horseback. He described these five years in Siberia as a 'genuine education in life and human character' (MR 168).

Arriving at Irkutsk, the capital of East Siberia, in the autumn of 1862, Kropotkin immediately became friends with General Kukel, to whom he was attached as aide-de-camp. An intelligent and practical man with radical views, Kukel came from Lithuania, and was not yet thirty-five years old. Kukel had been friendly with Bakunin during his exile - Bakunin had only recently, in 1861, escaped from Siberia - and it was from Kukel that Kropotkin first learned about the activities of this famous revolutionary (WA 54).

As Kukel had recently been appointed temporary governor of the province of TransBaikal, he invited Kropotkin to travel with him to its capital, Chita, a small frontier town around four hundred miles from Irkutsk.

Kropotkin was soon appointed secretary of two government committees: one to look into the reform of prisons and the system of exiles, the other to prepare a scheme for municipal self-government. Kropotkin, then only twenty years old, set to work on these reforming projects with great enthusiasm. He took the projects seriously: he read and studied books and reports on the subject of prison and local government, travelled widely, and met and discussed the issues with a considerable number of people.
But his efforts came to nothing, because following the Polish insurrection of 1863 a wave of reaction set in.

The revolution in Poland in January of 1863, as already mentioned, was brutally repressed. Thousands were killed in the conflict; many hundreds were hanged and scores of thousands transported to remote parts of Russia or Siberia, sentenced to hard labour or exile. Kropotkin met many of these Polish exiles, and quickly came to realise that no form of 'reform' was possible in the growing climate of suppression and reaction - even in remote Siberia. Seeing that there was nothing more that could be done in the way of reforms at Chita, Kropotkin decided to explore the Amur region.

His first journey was made in the summer of 1863 when he accompanied a flotilla of about thirty barges that were being floated down the Shilka and Amur rivers. These carried the annual supplies of flour, salt and cured meat to a number of pioneer settlements that stretched along the north bank of the Amur and its tributary the Usuri. Kropotkin acted as assistant to the chief of the flotilla, though he had no experience at all in river navigation, nor of the local Cossacks. He was much impressed with the landscape, the river flowing amidst mountains with steep wooded cliffs, offering 'some of the most delightful scenes in the world' (MR 189).

Unfortunately, however, some of the cargo was lost in a storm, and when Kropotkin returned to Irkutsk - travelling the two thousand miles back up river by rowing-boat, steamer and horseback - he was immediately requested to report to St Petersburg to explain the losses. This involved another journey by coach and rail of more than four thousand miles, taking almost a month. He soon discovered that the government civil servants in St Petersburg knew very little about conditions in Siberia.

Kropotkin did not stay long in St Petersburg. He returned to Irkutsk that same winter to take up a post as attaché to the governor-general of eastern Siberia, with responsibility for Cossack affairs. This meant residing at Irkutsk, but as he had little work to do, Kropotkin gladly accepted a proposal to undertake geographical exploration in Manchuria. So in 1864, disguised as a Russian merchant, he led an expedition there, taking with him forty horses and some cloth. He was accompanied by eleven Cossacks and one Tungus, all, including Kropotkin, travelling on horseback. They carried no firearms, apart from an old matchlock gun taken by the Tungus, which was used to procure game. The expedition began at Chita and crossed the Great Kingan mountain range to the ancient Chinese city of Merghen (Nen-Chiang), where they sold the horses at a profit. They then traversed uncharted territory to the Chinese frontier town of Aigun on the southern bank of the Amur. Then they crossed the river to Blagoveshchensk before returning to Irkutsk.

In the autumn of the same year, 1864, Kropotkin made a still more interesting trip. This time he went by steamer, exploring the river Sungari (Sung-Hua Chiang), a tributary of the Amur. This took him into the very heart of Manchuria, to the Chinese provincial capital of Ghirin (Chi-lin). This expedition was led by a Colonel Chernyaeff, and it included, besides Kropotkin, a doctor, an astronomer, two topographers and about twenty-five Cossack soldiers. The pretext for the trip was to convey a message of friendship to the governor-general at Ghirin. Kropotkin seems to have enjoyed visiting the local peasant villages and establishing cordial relations with the people. But he found the Chinese officials rather arrogant.

The following year, 1865, Kropotkin explored the Tunkirst Valley and the Sayan Highlands to the southwest of Lake Baikal. His final Siberian expedition was in 1866. Again there was a practical aim: to discover a direct means of communication between the gold mines of the Yakutsk province in Northern Siberia, on the Vitim and the Olokma Rivers, and Chita in the Transbaikal. This trip, through a wild, uninhabited region north of Lake Baikal, took three months during which Kropotkin covered over twelve hundred miles.

First he first visited the Lena goldmines and the Tikhono-Zadonsk mining settlement, which was later renamed Kropotkin after him, then, having fitted out the expedition, he travelled south over wild mountain ranges, following river gorges and wandering over alpine deserts and marshy plateaux. He was accompanied by a young naturalist, Ivan Poliakov, and an old Yakut hunter, who acted as guide, and carried an old birch-bark map which Kropotkin implicitly trusted.

Kropotkin later described this trip as 'a wonderful feat'. It also confirmed his theory that the structure of the mountains of Siberia ran southwest to northeast, and not parallel to the great rivers, that is, south to north, as indicated on the existing maps. (For more information on Kropotkin's Siberian travels, see MR 184-215, WA 61-74, Slatter 1981.)

These expeditions in Siberia offered Kropotkin much food for thought. He recorded his geographical and palaeontological observations and theories in a series of scientific papers and reports, published later in the *Geographical Journal* and in other academic journals. But his travels and experiences in Siberia had a profound impact on Kropotkin in many other ways.

Firstly, his contacts with Polish exiles, and his visits to state prisons and to the Lena mining settlements, gave Kropotkin firsthand experience of the realities of nascent capitalism and the brutality of the Tsarist state. His visit to the lena goldfields, in particular, had a disturbing impact on Kropotkin, for he discovered that the workers there toiled from 4am in the morning till 8pm at night, in all weather conditions, and for a very meagre wage. In June 1866 he wrote to his brother Alexander about the Lena mines: 'This is where one can gaze every day to one's hearts content upon the enslavement of the worker by capital'. To Kropotkin these conditions seemed even more brutal and soul-destroying than the worst excesses of serfdom (MK 68-69).

He was equally incensed at the harsh treatment meted out to Polish exiles transported to Eastern Siberia after the 1863 insurrection. These exiles were subject to 'hard labour' in the salt mines and iron works, or worked as navvies building roads. A group of exiles that attempted to escape in the winter of 1866 was summarily executed by hanging. Kropotkin's brother Alexander had been posted to Siberia in 1864 to command a squadron of the Irkutsk Cossacks, to the delight of Kropotkin, for he missed his brother's warm friendship and intellectual stimulus. Alexander soon married Vera Chaikovskaia, the daughter of a Polish exile family that he had befriended. But the execution of the Polish exiles who tried to escape, and the brutal conditions under which exiles worked in Siberia, made Kropotkin and his brother painfully aware of what it meant to belong to the army. They therefore decided early in 1867 to leave military service and to return to Russia (MR 223).

A second major influence on Kropotkin from his time in Siberia was that, through his contact with Kukel and various political exiles in eastern Siberia, he came to meet the famous Russian poet Mikhail Mikhailov. The poet had been condemned to hard labour in 1861 for issuing a revolutionary tract. Suffering from tuberculosis, Mikhailov died in 1865, although local officials, sympathetic both to his political ideas and his condition, did what they could to alleviate his suffering. It was through Mikhailov that Kropotkin first became acquainted with anarchist ideas, for the poet gave him an annotated copy of Proudhon's *Le Système des Contradictions Économiques ou la Philosophie de la Misère* which Kropotkin read avidly (Nettlau 1921, WA 57).

Thirdly, Kropotkin's frustrations in attempting to carry out necessary reforms in relation to prisons and local government, and his direct experiences with Siberian peasants, boatmen, Cossacks and local tribal people (Tungus, Buryat) made Kropotkin realise that little worthwhile could be achieved through the machinery of the state. As a member of a landed aristocracy he had been brought up to believe in the necessity and efficacy of hierarchy and in the 'principle of command', and had a great deal of confidence in 'the necessity of commanding, ordering, scolding, punishing and the like'. But he soon realised that not only did ordinary native people have complex forms of social organisation that did not entail state power, but that the "principle of common understanding" was much more effective in organising matters of 'real life' than that of 'command'. He writes that in Siberia he lost 'whatever faith in state discipline I had cherished before'. He was now prepared to become an anarchist (MR 216-17).

A further consequence of Kropotkin's travels in Siberia was that, while undertaking zoological, ethnographic and geographical researches in Siberia, he came to doubt Darwin's theory of evolution, in the sense emphasised by his disciple Thomas Huxley, that this entailed simply a competitive struggle for existence. Kropotkin became aware that reciprocal relations and 'mutual aid' were significant aspects of biological existence, as well as of social life. He would later develop these ideas in his famous work, *Mutual Aid*, published in 1902.

Kropotkin's geographical, zoological and ethnographic researches in Siberia showed that he was a penetrating and acute observer, both of the natural world and of social life. But his experiences in Siberia also led to his complete disillusionment with the Tsarist state. As Martin Miller writes:

> He went to Siberia in 1862 full of enthusiasm for the possibilities of national reform. He left five years later completely disillusioned (MK 70).

But it was through his disillusionment, and because of his close identification with the natural world: the poetry of nature, that Kropotkin came to forge his own unique philosophy of life and came to affirm and develop his own political credo - anarchist communism. In this he contrasted with his brother Alexander who increasingly experienced despair and lacked Kropotkin's optimism about the future.

Kropotkin, along with his brother, returned to St Petersburg early in 1867. The next five years of his life were almost exclusively devoted to scientific work, mainly in relation to his geographical researches and to his studies at the university. He entered university specifically to get a thorough training in mathematics and physics. He was nevertheless anxious, though wishing to be independent and to give up military service, not to distress his father unduly.

> My brother entered the military academy for jurisprudence while I entirely gave up military service to the great dissatisfaction of my father, who hated the very sight of civilian dress. In order not to completely distress my father, I transferred to civil service (MR 224).

Kropotkin obtained a post in St Petersburg with the Ministry of Internal Affairs - a position that required little attention - while at the same time engaging in his geographical studies, and in writing up his Siberian researches. His studies were greatly enhanced by his appointment as secretary of the Physical Geography Section of the Russian Geographical Society. Through the society, Kropotkin was able to meet and exchange ideas with many leading geographers and naturalists: in his memoirs he mentions such scholars as Syevertov, Fedchenko and Przemalsky, all of whom were well-travelled and had extensive zoological interests (MR 228-230). Also through the Geographical Society, Kropotkin developed an interest in Arctic exploration: he made what he described as a 'modest tour' of Finland in 1871, specifically to study glacial deposits. At this time, as already noted, he enrolled as a university student in the Faculty of Mathematics and Natural Sciences in St Petersburg, but he never did complete his degree. In order to support himself, he began translating the works of Herbert Spencer into Russian - though by all accounts Kropotkin himself at this time lived a simple and frugal life.

Living in St Petersburg, Kropotkin became increasingly aware of the deteriorating political situation in Russia. On joining some of the literary circles, he soon realised that people tended to avoid any discussion of politics. The more radical that people had been in the early part of Alexander II's reign, the more they feared to admit their views. Since the young student Dimitri Karakozov had shot at Alexander II in April 1866 (he was hanged for his foolhardy attempt to rid Russia of autocracy), the state police had become virtually omnipotent. The policies of Alexander II had assumed a decidedly repressive and reactionary character, and any one suspected of radical, or 'dangerous' views - and this might mean anything - was liable to be arrested. Such people might end up spending years secluded in the fortress of St Peter and St Paul; be subject to torture; exiled, or sentenced to hard labour in Siberia or in some remote part of Russia. At the theatre, Kropotkin noted, the radical Italian operas had given way to the light-hearted musicals of Offenbach (MR 252).

The only ray of light in this gloomy political scene was, for Kropotkin, the emergence of the feminist movement in Russia. Aristocratic women were beginning to obtain access to higher education and to the professions. But this 'grand movement', he recalled, was not merely feminist: the desire simply to get a share of privileged positions, but also had a more radical import, in that the women's sympathies embraced ordinary people. He noted that Tsar Alexander II had an open hatred for educated women, and actuallly trembled when he saw a woman wearing spectacles and a Garibaldi cap, fearing she might be a nihilist bent on shooting him (MR 258-63).

The years 1870-71 were particularly eventful and significant for Kropotkin. His travels and experiences in Siberia had developed in him a sense of independence and an understanding of humans from a wide variety of backgrounds and circumstances. While on his travels, engaged in geographical, ethnographic and zoological researches, he had also developed an interest in the scientific method, and a feeling for science that he would retain all his life. But when, in 1871, Kropotkin was offered the post of secretary to the Russian Geographical Society, he decided, after much reflection, to decline the offer. He was resolved not to devote his life purely to scholarship, however attractive this might be. Although geographical research might lead to the 'highest joys' it would, he felt, be at the expense of working people, and lead him to neglect the needs and aspirations of the peasants, of the 'poor and down trodden'. The only way that he could justify his own life, Kropotkin reflected, was to devote himself to the liberation of the masses. He concluded:

> This is the direction in which, and these are the kind of people for whom, I must work (MR 240).

Two other events in 1871 were also to have an important influence on Kropotkin's change of direction: the Paris Commune and the death of his father. The outbreak of the Franco-Prussian war and, more particularly, the rise and downfall of the Paris Commune of 1871, had a profound impact on Russian society. Kropotkin's friend Dimitri Klements claimed that the Paris Commune was a 'turning point' in the Russian revolutionary movement, and throughout the summer of 1871 Kropotkin's letters are full of references to the Commune (MK 74). He was later to write that he never suffered so much as when he read the book *Le Livre Rouge de la Justice Rurale*, which contained newspaper reports on the last days of the Paris Commune in May 1871, and detailed the brutal repression of the insurgents. As he read those pages, Kropotkin recalled, he was seized with a 'profound despair of mankind' (MR 285).

The death of Kropotkin's father took place in the autumn of 1871. The latter had not been well for some time, and Kropotkin describes in his memoirs his last meeting with him in the spring. When they parted, his father 'seemed almost to dread his gloomy loneliness amidst the wreckage of a system he had lived to maintain'. Kropotkin was in Finland at the time of his father's death, and hurried to Moscow, only to arrive when the burial service was already in progress. His father was buried in the same old red church in Moscow where he had been baptised (MR 265).

On his father's death Kropotkin found himself not only free of filial obligations, but also wealthy, as he had inherited one of his father's estates, Tambov. But he had no intention of becoming an idle aristocrat, living off the income from his land like many of his contemporaries. His thoughts turned instead to finding a way in which he could devote himself to the oppressed people of Russia. Influenced by Sergei Kravchinsky, better known as Stepniak, and by the sister of his brother's wife, Sofia Lavrova, both of whom were already committed revolutionary socialists, Kropotkin decided to visit Switzerland, then the Mecca of radical thought.

3: Becoming an Anarchist

It has been suggested that the circumstances of Kropotkin's early upbringing - his father's harsh neglect and authoritarian outlook; his mother's gentle intelligence; his own personal contacts with the peasants on the Nikolskoye estate, and the radical influence of his tutors - all contributed towards the fostering of his independence of thought, and the compassion he felt towards the peasants and working people. His experiences and contacts during his travels in Siberia only affirmed his sense of rebellion against authority, and brought him into contact with radical ideas and literature. The works of Herzen, Proudhon and Chernyshevsky (whose famous, but rather shallow novel *What is to be done* was published in 1863), were important to Kropotkin, and it was no surprise that, like Tolstoy, he enthusiastically supported the emancipation of the serfs. By the end of his sojourn in Siberia he had become firmly convinced of the brutal and entirely reactionary nature of the Tsarist autocracy, and no longer expected any improvements to come by way of government 'reform' (WA 93). Like many of his aristocratic contemporaries, Kropotkin also felt an impelling need to devote his life to alleviating the hardships and sufferings of the common people:

> that desire to expiate a sense of social guilt by devoting himself to the most down-trodden members of society, which was such a distinctive feature of the Narodnik intelligentsia in his age (WA 94-95).

But in following the path of the Narodniks, Kropotkin never expressed the kind of guilt-ridden sentiments that one associates with Tolstoy. Because he had an intimate knowledge and understanding of peasant life he never came to idealise the peasantry. Still less was he ever an advocate of individual terrorism. But Kropotkin's first journey to Western Europe in the spring of 1872 was a crucial event in his life, and the political contacts he made with the Jura Federation are seen as heralding his 'conversion' to a revolutionary career.

Kropotkin arrived in Zurich in February 1872, and immediately went to the house of his brother's sister-in-law Sofia Lavrova, who was studying at the University. At that time Zurich was full of Russian students, both men and women, mostly of a radical persuasion. Sofia had become a member of the International Workingmen's Association, set up in London in September 1864, and was particularly associated with its Bakuninist faction. The Association had begun, as G. D. H. Cole put it, primarily as a 'Trade Union affair', expressing the solidarity of workers in France and Britain (1954: 88). Most of the French participants at the 1864 proceedings were, however, not industrial workers but artisans, who were essentially followers of Proudhoun's kind of socialism: mutualism.

It has to be remembered that the 'First International' was not the creation of Marx, nor was it specifically Marxist at its inception. The movement spread like wildfire through France, Italy, Spain and Switzerland, for it represented the idea of an international association of working people, who were called upon to unite – 'without distinction of creed, sex, nationality, race or colour' - in the struggle against capitalism. It proclaimed that the emancipation of the workers must be the task and the act of the workers themselves (MR 271).

In July 1868 Michael Bakunin had become a member of the Geneva section of the International Workingmen's Association, for many of his associates were already members. This eventually led to the famous split in the International, although the idea put out by Marx that Bakunin was simply a political intriguer who was out to 'wreck' the Association, was quite false. Bakunin, in fact, did far more than Marx to expand the membership of the International (Morris 1993: 59).

The conflict between Marx and Bakunin came to a head at the sham conference of the International held in London in September 1871. This conference affirmed the authority of the general council (under Marx); declared the necessity for workers of each country to form their own political party, and disparaged anarchism as a political heresy. The Swiss groups of the International, almost all followers of Bakunin and thus hostile to Marx, immediately organised their own conference at Sonvilier in the Jura, in November 1871. They denounced the autocratic powers exercised by the general council, and called for the re-affirmation of an International movement that was composed of a free federation of autonomous sections instead of one governed by a general council. The congress produced the 'Sonvilier Circular', which criticised the idea of the 'conquest of political power by the working class'.

The split of the International crystallised around the leading figures of the organisation, Karl Marx and Michael Bakunin. Woodcock and Avakumovic express their differences succinctly:

> The two men were as different in character as in ideas. Marx, the bitter, dictatorial scholar with a great power of social analysis that had been submerged in a messianic conception of history; Bakunin, the hero of insurrections and prisons, the generous and able orator, extravagant in his enthusiasm, too impatient to be a systematic thinker, but possessed of a political clairvoyance that enabled him to see with remarkable accuracy the defects of his opponents and their teachings. But it was much more than a struggle of personalities, it was also a clash of two wholly

different conceptions of social organisation, two mutually alien philosophies of life (WA lll).

These were, respectively: the state socialism of the Marxists, which put an emphasis on authority, and acknowledged the need for a revolutionary government - the dictatorship of the proletariat - to secure the development of communism; and Bakunin's collectivism, which advocated the abolition of the state, and its replacement by a stateless form of communism: a federal society based on free communes and voluntary associations. Marx was all in favour of the conquest of political power; Bakunin advocated its dissolution, and favoured direct economic action.

For many years Kropotkin had longed to learn about the International Workingmen's Association, and so, as soon as he had reached Zurich, and found his brother's sister-in-law, he joined it, through her. He also met another close disciple of Bakunin, Michael Sazhin. Sofia brought Kropotkin piles of pamphlets, books and newspapers relating to socialism and these he studied avidly, spending days and nights reading. He also subjected Sofia and Sazhin to long sessions of persistent and detailed questioning about the International and the Paris Commune. Such discussions often became intense and very heated. Kropotkin recalled in his memoirs that his reading of the socialist literature opened up a 'new world' to him, one previously unknown both to himself and to current sociological theory (MR 274-5, MK 78).

Kropotkin was anxious to meet ordinary working people, and especially to encounter workers in their everyday life. So, armed with letters of introduction from Sofia, he went on to Geneva, then the centre of International Socialism. Switzerland was not only the meeting place of many Russian exiles, but had also become a place of refuge for many French socialists known to have been involved in the Paris Commune, who had to flee the country when reaction set in. In Geneva Kropotkin met Nicholas Utin and Olga Levashova, both Russian exiles. They were local leaders of the International, and through them Kropotkin met other members of the Association. The Geneva Section met in a large hall at the Masonic temple, where they held lectures and political meetings. Kropotkin writes that: 'It was a people's university as well as a people's forum' (MR 276).

Kropotkin initially got on well with Utin, whom he describes as a 'bright, clear and active man'. Utin had been exiled from Russia since 1863, and had at one time been a close friend of Bakunin. But as the split developed between the Marxists and Bakunin's Alliance of Social Democracy, Utin sided with the state socialists and parliamentarians. He seems to have been a person of dubious sincerity, for in order to curry favour with the 'venerable Marx', he spread rumours that Bakunin was a Tsarist agent, doing much to poison the relationship between the two men. Significantly, after having helped to undermine the International, Utin made peace with Tsardom, returned to Russia and ended his days as a wealthy and respectable government contractor (Cole 1954: 197).

But Kropotkin preferred to be with the workers themselves, in order to get a real understanding of the movement from the inside. He wrote:

> Taking a glass of sour wine at one of the tables in the hall, I used to sit there every evening amid the workers, and soon became friendly with several of them, especially with a stone-mason from Alsace, who had left France after the insurrection of the Commune (MR 276).

Kropotkin was much impressed by the integrity and devotion of the working people he met, many of whom led Spartan lives, refraining from both alcohol and tobacco. But he was not impressed by the political machinations of the leaders of the international in Geneva. He records the 'wire-pulling' and dishonestly of Utin, who made efforts to assist the parliamentary ambitions of a local lawyer (MR 278-80). Kropotkin angrily reproached Utin for not supporting the workers, who were then preparing to strike for higher wages.

Kropotkin could not reconcile the machinations of the leaders with their fiery speeches from the platform, and felt quite disillusioned. So he expressed to Utin his desire to become acquainted with the other 'Section' of the International. Its members were then known as the 'Bakuninists' or 'federalists', for at that time the term 'anarchist' was rarely used, and the split in the International between the state socialists (social democrats) and the libertarian socialists (anarchists), though evident, had not become a clear-cut ideological division. Utin, who remained friendly with Kropotkin, gave him a letter of introduction to another Russian exile, Nicholas Zhukovsky. With a sigh, and some foresight, Utin remarked to Kropotkin: 'Well, you won't return to us, you will remain with them' (MR 280).

Zhukovsky was a Russian aristocrat who had been in exile since 1862, and had served on the editorial board of many radical newspapers. From early on he had come under the influence of Bakunin, and was to remain an anarchist all his life. On meeting Kropotkin, Zhukovsky advised him to visit the real centre of the Bakuninist movement: the Jura Federation in the Jura mountains. Kropotkin went to Neuchatel, and made his first contacts with members of the Jura Federation, an association which was to play an important part in the development of socialism. He noted:

In 1872 the Jura Federation was becoming a rebel against the authority of the general council of the International Workingmen's Association. The association was essentially a workingmen's movement, the workers understanding it as such and not as a political party (MR 281).

But when in 1871, at a secret conference in London, the general council, supported by a few delegates, had directed the Association to support electoral agitation, this engendered strong opposition from the Jura Federation. This opposition represented, Kropotkin wrote, 'the first spark of anarchism' (MR 282).

In contrast to what he had experienced in Geneva, there was no separation between leaders and workers in the Jura Federation: 'their leaders were simply their more active comrades - initiators rather than leaders' (MR 286). Among the more important of these was James Guillaume (1844-1916), a former philosophy teacher at the University of Zurich, who was then the manager of a small printing office in Neuchatel. Kropotkin described him as 'one of the most intelligent and broadly educated men I have ever met' (MR 282). Although never intimate, the two men were to remain close friends and comrades throughout life. Along with Bakunin, Guillaume had been one of the key figures in opposition to Marx's authoritarian communism.

The Jura Federation held its first separate congress at Sonvielier in November 1871, as noted above. At this meeting Guillaume vigorously protested against Marx's attempt to establish an 'authoritarian structure" within the International. He wrote a long paper: 'Le Collectivisme de l'Internationale', advocating federalism and the principle of autonomy for all Sections. Federalism, reflecting the libertarian, anti-statist perspective of Bakunin, thus became adopted by the Jura Federation. Kropotkin, strongly identifying with the independent

Jura workers, most of whom were engaged in the watch-trade, came to acknowledge the libertarian socialism that Guillaume, following Bakunin, advocated. He also went to the mountain village of Sonvilier and there met the anarchist watchmaker Adhemar Schwitzguebel, who was a close friend of Bakunin. Afterwards the two men became close associates.

While in Switzerland Kropotkin also came to meet many of the leading Communards: Gustave Lefrançois, Louise Michel, Elisée Reclus and Benoît Malon. From Malon, who later became a founder of the independent socialists in France, and wrote an important *History of Socialism*, Kropotkin learnt in graphic detail the history of the Paris Commune, its heroism and its mistakes, and the moral integrity and suffering of the defeated revolutionaries. Kropotkin recalled how every day he went to see Malon, to hear what this quiet, poetical and good-hearted Communard had to tell him about the Commune, in which he took a prominent part (MR 284).

Kropotkin summed up his experience in the Jura Mountains like this:

> The theoretical aspects of anarchism, as they were then beginning to be expressed in the Jura Federations especially by Bakunin; the criticisms of state socialism - the fear of an economic despotism, far more dangerous than merely political despotism - which I heard formulated there and the revolutionary character of the agitation, appealed strongly to my mind. But the equalitarian relations which I found in the Jura Mountains, the independence of thought and expression which I saw developing in the workers, and their unlimited devotion to the cause appealed far more strongly to my feelings; and when I came away from the mountains, after a week's stay with the

watchmakers, my views upon socialism were settled. I was an anarchist (MR 287).

But surprisingly, although Bakunin was in Locarno at the time, Kropotkin never came to meet the famous Russian revolutionary. This he very much regretted, because when he returned to Switzerland four years later, Bakunin had recently died, on July 1st 1876, some two weeks after Kropotkin's dramatic escape from prison. There has been some speculation as to why the two anarchists never met. It has been suggested that Bakunin did not wish to meet Kropotkin because he associated Kropotkin with his brother Alexander, who was a disciple of Bakunin's populist rival, Lavrov, and that he was offended that Kropotkin had spent so much time with the intriguer and sycophant of Marx, Utin, rather than with Zhukovsky.

But Kropotkin later suggested to the anarchist historian Max Nettlau that Guillaume had advised him not to visit Bakunin, because the old revolutionary was feeling frail and overwrought by the recent struggles with the Marxist section of the International (WA 121). Most of the people Kropotkin met in Switzerland: his brother's sister-in-law Sofia, Sazhin, Zhukovsky, Guillaume, Schwitzguebel and Malon, were either friends or disciples of Bakunin, and he was most impressed that in their conversations they always spoke of Bakunin not as a 'leader' but as a personal friend. He was a person of whom everyone spoke with love and in a spirit of comradeship. What struck me most, Kropotkin wrote, 'was that Bakunin's influence was felt much less as the influence of an intellectual authority than as the influence of a moral personality' (MR 288).

After undertaking a short visit to Brussels, Kropotkin returned to St Petersburg in May 1872. He carried with him a whole trunkload of socialist books and newspapers, all 'unconditionally prohibited' by Russian censorship. He therefore decided to return to Russia from the South, via Cracow, and elicited the help of some Jewish smugglers to get his large collection of socialist literature across the Russian frontier. He arrived in St Petersburg to proudly display his 'trophies' to his brother Alexander. Kropotkin recorded that he returned from Switzerland 'with distinct sociological ideas which I have retained since, doing my best to develop them in more and more definite concrete forms' (MR 289).

On his arrival back in St Petersburg Kropotkin immediately joined the circle of Chaikovsky, a nihilist group with whom he was closely associated for two years and which left an indelible impression upon all his subsequent life and thought.

4: Russian Populism and the Chaikovsky Circle

Much has been written both on nihilism and on Russian populism. These were widespread radical movements that erupted in Russia in the middle of the nineteenth century. The two movements were closely interlinked, and were born during the same social and intellectual upheavals that followed the death of Tsar Nicholas I in 1855. Both movements virtually came to an end in November 1881 with the assassination of Alexander II by a group of radicals known as 'The People's Will' (*Narodnaya Volya*).

Nihilism, as Kropotkin insists, was ill-understood in Western Europe, for it was continually equated with terrorism. This he felt was a mistake. To conflate nihilism with terrorism, he wrote, was akin to confusing a philosophical movement like Stoicism or Positivism with a political doctrine like Republicanism (MR 297). Nihilism was less a political grouping than a philosophical or cultural movement that called for the complete transformation of social attitudes and conventions. Nihilism was the name Turgenev gave it in his famous novel *Fathers and Sons* (1862); as a movement it was particularly associated with the literary critic Dimitri Pisarev (1841- 68). A radical and a feminist, Pisarev spent four years of his active life in solitary confinement, and died at the early age of twenty-seven in a drowning accident. He never called himself a 'nihilist'.

Originally the term nihilism simply meant the radical rejection of established conventions and institutions, and an attitude that recognised nothing (*nihil*) that could not be justified by rational argument (Walicki 1980: 210). In its cultural aspect, nihilism declared war upon the 'conventional lies of civilised mankind', as Kropotkin described it: the religious superstitions, the prejudices, the hypocrisy, the sham sentimentality, the art for art's sake, the utter insincerity of the Russian aristocracy which the young revolutionary radicals saw as exemplified in the generation of their fathers (MR 278-79). To emphasise their radical rejection of aristocratic conventions, the young nihilists wore blue-tinted spectacles and high boots, smoked cigarettes and generally touted an unkempt appearance. The men wore dishevelled beards and let their hair grow long, while the women, who played an important part in the 'nihilist protest', cropped their hair, abandoned combs and jewellery, and wore plain clothes (Broido 1978: 18). The nihilists were also critical of family conventions – they advocated gender equality and sexual freedom - and of the Russian Orthodox Church, as well as of the Tsarist state.

In his classic study on the origins of Russian communism, Nicholas Berdyaev (1960) emphasised that, as an intellectual movement, nihilism was as narrow and dogmatic as the Orthodox Russian Church. For in its emphasis on individualism: 'rational egoism', and on materialism and science (positivism) and on utilitarianism (the nihilists eclectically gathered ideas from many sources: Comte, Saint-Simon, Mill, Fourier, Spencer), the Russian prophets of the Enlightenment, that is, the nihilists, completely lacked the open, sceptical and truly scientific temper of men like Voltaire and Diderot. Nihilism, wrote Berdyaev, was 'orthodox asceticism turned inside out'; its materialism constituted a particular kind of dogmatic theology which was later characteristic of the Bolsheviks.

The attitude of Russian nihilists to science, Berdyaev argued, was 'idolatrous', dogmatic science becoming an article of faith. In fact, the nihilist prophets of the Enlightenment were not men of science, and as a movement, he concluded, nihilism could be viewed as a distinctively religious phenomenon (Berdyaev 1960: 46-48). The positivist and materialist faith of the nihilists thus bordered on fanaticism, while they emphasised 'rational egoism' to such a degree that it implied not socialism, but rather a theory that 'exalted economic calculation and utilitarian coldness' (Venturi 1960: 327).

More extreme nihilists often resorted to terrorism. In 1862, more than thirty years after the Decembrist uprising, the first stirrings of a revolutionary movement in Russia became evident. It was manifested in the formation of a secret revolutionary party which called itself *Zemlya I Volya* (Land and Freedom). Although almost immediately suppressed by the Tsarist secret police, acts of terrorism continued to erupt. The St Petersburg fire of 1862, which Kropotkin had experienced while in the Corps, was thought to have been an act of arson. In 1866 the young student Dimitri Karakozov attempted to assassinate the Tsar.

The nihilist conspirator who was to achieve most notoriety was Sergei Nechaev (1847-82). In his famous 'Revolutionary Catechism' (1869), Nechaev had written: 'Karakozov was merely the prologue, now friends, let us start the drama'. Of peasant background, Nechaev had been the instigator of a number of conspiratorial and revolutionary groups and had become a close friend of Bakunin, until the anarchist came to recognise, around 1870, that Nechaev was a violent, rather disturbed and unscrupulous character, even though full of revolutionary fervour. Captured in Switzerland in August 1872, Nechaev was taken back to Russia. He was imprisoned for ten years in the St Peter and St Paul fortress, in solitary confinement, and there died of scurvy at the age of thirty-five (Morris 1993: 46).

It has been suggested that Kropotkin had many nihilist characteristics, in that he combined a belief in freedom with an essential puritanism, and that he had an 'almost religious veneration of science'. But this portrait is somewhat overdrawn; Kropotkin repudiated egoism, emphasising the culture and creativity of every individual, while art and poetry was always an intrinsic part of his philosophical outlook. There was, moreover, 'a pantheistic emotion in his love for nature' (WA 101-2). In spite of his reverence for science, it is quite misleading to interpret Kroptkin as a 'positivist', for he was always critical of instrumental rationalism and the utilitarian outlook of such 'nihilists'. He was also, like Bakunin, highly critical of the amoral cult of violence and the Jesuit methods that the embittered young Nechaev seems to have embraced.

Around 1870 there was a decided shift of emphasis in the Russian radical ethos, away from nihilism and towards populism. Russian populism (*Narodnichestvo*) was not a political party, nor in any sense a coherent body of doctrine, but rather a loose political movement that essentially embraced a form of agrarian socialism. Almost all the revolutionary intellectuals that came into prominence in Russia during the 1860s: Herzen, Bakunin, Chernyshevsky, Lavrov -were in essence *narodniks* (populists), in that they advocated 'going to the people' (*Narod*, people), and suggested a form of socialism based on the village commune.

The one important revolutionary socialist who was an exception and cannot be considered a populist was Peter Tkachev (1844-86). A devoted follower of Karl Marx, Tkachev advocated, like the Jacobins, the seizure of political power by a centralised and disciplined body of professional revolutionaries. Although a former associate of Nechaev, he did not however advocate individual terrorism. But he was critical of Bakunin and the populists; he thought ordinary people were incapable of liberating themselves by their own efforts, and he has been rightly seen as a precursor of Lenin and the Bolsheviks.

The populist doctrine was expressed in a number of important texts: in Chernyshevsky's novel *What is to be done* (1863) and in Lavrov's *Historical Letters* (1870), as well as in the important writings of the radical liberal Alexander Herzen (1812-70), who was an exile in London for much of his later life. There he had edited the first Russian emigré newspaper *The Bell* (*Kolokol*) which was smuggled into Russian during the years

1857-67. At that time it was heralded as the voice of free Russia. Franco Venturi wrote that:

> Herzen created populism ; Chernyshevsky was its politician (1960: 29), for it was Nicholas Chernyshevsky (1828-89), a self-taught scholar and a follower of the French utopian socialist Fourier, who provided populism with its basic content, in his advocacy of agrarian socialism and in his defence of the peasant commune.

Isaiah Berlin (1978) has lucidly sketched some of the main features of Russian populism. The central aims of the movement were social justice and social equality, and, following Herzen, populists felt that the essence of a just and equal society already existed in the Russian agrarian community, the *Obshchina*, organised through the *Mir*, a free association of peasants who periodically redistributed land. The populists were highly critical of industrial capitalism, and shared the democratic ideals of the European intellectuals of their day. But their socialist politics were more akin to those of Fourier and Proudhon rather than of Marx. As Berlin writes:

> The populists believed that the development of large-scale centralised industry was not 'natural' and therefore led inexorably to the degradation and dehumanisation of all those who were caught in its tentacles; capitalism was an appalling evil, destructive of body and soul.

The populists denied that social and economic progress was necessarily bound up with industrial capitalism:

They maintained that the application of scientific truths and methods to social and individual problems (in which they passionately believed)... could be realised without this fatal sacrifice (Berlin 1978: 212).

The Russian populists were not mystical nationalists, and they saw no reason why Russia should not benefit from western science and technology. But they thought it possible to avoid both the evils of capitalism and the despotism of the centralised government by advocating a form of political structure that consisted of a loose federation of self-governing units of production, as suggested by Fourier and Proudhon. Berlin writes that:

All Russian populists were agreed that the state was the embodiment of a system of coercion and inequality, and therefore intrinsically evil: neither justice nor happiness was possible until it was eliminated (op. cit. 217).

This is not in fact the case. Although faith in human freedom, and an emphasis on the individual and reason, was an important aspect of Russian populism, which also advocated in general a libertarian form of agrarian socialism, populism was not anarchism, as Berlin concedes. Neither Chernyshevsky nor Lavrov were anarchists. Chernyshevsky, though believing in the need to preserve the peasant commune and advocating a form of socialism that avoided the agonies of industrialism, nonetheless upheld the necessity of the state. This 'prodding genius', as Berlin describes him, regarded rational egoism as a logical outcome of materialism, and though critical of the Tsarist absolutist state, Chernyshevsky essentially pleaded, like Herzen, for a radical and 'enlightened' bourgeois democracy (Walicki 1980: 194-202). As

Berlin notes, Chernyshevsky was the least anarchistic of the populists, and in essence a state socialist.

Peter Lavrov (1823-1900), too, was essentially a social democrat. As we have noted, Kropotkin's brother Alexander was a close friend and disciple of Lavrov. Although a leading populist, and by all accounts an attractive person, Lavrov was strongly influenced by Marx. He therefore came to emphasise a crucial role for the state in the transition to agrarian socialism. Lavrov later became an exile and took an active part in the Paris Commune. From 1873 he published in Geneva the radical journal *Forward!*, which for many years was the chief organ of the populist movement in Russia (Utechin 1963: 129).

The important difference in the political views of Kropotkin and his brother Alexander is that Kropotkin was essentially an activist and a follower of Bakunin, while Alexander was more philosophical and less involved with popular rebellion, and identified with Lavrov's social democratic politics (MK 84). Indeed the political differences between Kropotkin and his brother reflect the two tendencies within the populist movement: the one oriented towards Bakunin with its emphasis on the common people, the other towards Lavrov with an emphasis on the radical intelligentsia. But both advocated 'going to the people' (WA 127). (Important studies of Russian populism and its revolutionary tradition include Yarmolinsky 1957, Venturi 1960 and Walicki 1980.)

Like Herzen, Lavrov and Turgenev, Kropotkin was brought up in the cosmopolitan atmosphere of the Russian aristocracy; he spoke and wrote French fluently, and had a deep interest in scientific scholarship. He therefore tended to think internationally, and was free of any narrow chauvinism (WA 107). But having spent the spring of 1872 with the most active sections of the International, Kropotkin, on his return to Russia, immediately became involved in the populist movement that was then coming into prominence. Having become friendly while at university with Dimitri Klements (called Kelnitz in Kropotkin's memoirs), Kropotkin was invited by Klements to join a radical populist group known as the Chaikovsky Circle. Franco Venturi, in his excellent study of Russian populism (1960), described the Circle as the first large populist movement. Indeed he suggests that populism only really began around 1870, when various groups, despairing of the kind of insurrectionary tactics advocated by Nechaev, put an emphasis on 'going to the people' (cf *Narod*, mentioned above).

The real founder of the Chaikovsky Circle was a medical student, Mark Natanson (1850-1919), who in 1869 had gathered around him a group of young radicals at the School of Medicine in St Petersburg. Natanson was a devoted admirer of Chernyshevsky and evidently a young man of great energy and enterprise. Early in 1871 he was arrested and exiled to Archangel. He was later to become one of the founders, along with Georgii Plekhanov, of the second *Zemlya i Volya* (land and freedom) revolutionary group, and would eventually come to be associated with the socialist revolutionary party in the years prior to the Russian Revolution.

After Natanson's arrest, Nikolai Chaikovsky (1850-1926), who lent his name to the group, became convenor of the Chaikovsky Circle. He was only twenty years old but he made a great impression on Kropotkin, nearly ten years his senior, for he was a strong, dependable character, and an able organiser. He was the son of a provincial civil servant, a brother of the famous composer, and had mystical leanings, but he was to remain a life-long friend of Kropotkin. Like Natanson, in his later years he actively supported the Socialist Revolutionary Party.

The Chaikovsky Circle was initially a study group, united for the purposes of self-education. But soon after Kropotkin joined in May 1872, it took on a more revolutionary character, and began distributing radical and socialist propaganda among the peasants, as well as among the workers of St Petersburg. The Circle was in the nature of a closely-knit fraternity - a united group of friends, as Kroptkin described it - and the Circle accepted as members only people whom they knew they could trust. There were no initiation rites nor oaths of allegiance, and although the group was conspiratorial, as even distributing socialist literature in Tsarist Russia was viewed as a serious offence against the state, the circle repudiated terrorist forms of action (MR 303-6). The group was initially suspicious of Kropotkin, because he came from the highest echelons of the Russian aristocracy, had an elite military background, and was much older than the rest of the Circle. But he was eventually accepted, Sophia Perovskaya suggesting to Natanson that he was dependable and 'young in spirit' (MK 90).

Kropotkin affirms that the Circle was not an organised movement, and in St Petersburg its membership was estimated at only thirty active members, although there were groups or Chaikovskists in Moscow, Odessa, Tula and Kazan. The circle rejected constitutionalism, and the emphasis was on agitation among the peasants and urban workers, and in being involved in propaganda they called *knizhnoe delo*, 'the cause of the book'. In going to the people and in advocating agrarian socialism, the Chaikovsky Circle was prototypically a *narodnik* (populist) movement. Furthermore, 'the circle was determined not to repeat the failures of the past, represented by the centralised, elitist conspiracy of Nechaev on the one hand, and the futile, individual terrorism of Karakozov on the other' (MK 91). Although all members of the Circle were populists, only Kropotkin, it seems, had anarchist tendencies.

In his memoirs Kropotkin writes in glowing terms about the two years that he spent working with the Circle of Chaikovsky. He noted that it left a deep impression upon all his subsequent life and thought: 'Never did I meet elsewhere such a collection of morally superior men and women as the score of persons whose acquaintance I made at the first meeting of the circle of Chaikovsky. I still feel proud of having been received into that family' (MR 306).

Many members of the Circle later became famous through their revolutionary activities; Kropotkin retained some as close friends throughout his life. There was Sophia Perovskaya, born in 1853, whose father had been military governor of St Petersburg. She was a passionate revolutionary and populist. As a person, Kropotkin describes her as generous, intelligent, loving and courageous. She was hanged in April 1881 for her part in her assassination of Alexander II.

There was Dimitri Klements, the person who had introduced Kropotkin to the Circle. Born in 1848, Klements came from a small land-owning family in South Russia: in his youth he was deeply influenced by the writings of Herzen and Chernyshevsky. He was a poet, a writer and a gifted scholar. Though politically something of a loner, he devoted his life to struggling against the Tsarist regime, and would spend time both in exile and in prison. He had travelled widely in Central Asia and was, like Kropotkin, a talented geographer. He remained a close friend throughout his life. In his memoirs Kropotkin pays tribute to his admirable personality (MR 303-4).

Finally, there is Sergei Kravchinsky (*nom de plume*: Stepniak) (1851-1895) who, like Kropotkin, began his extraordinary revolutionary career in the Circle of Chaikovsky. He later joined the *Zemlya i Volya* group, and was involved in the assassination of the army chief Mezentsev in August 1878. He was a man of great courage, and an excellent writer, and he became well-known for his profile of Russian revolutionaries, published in English as *Underground Russia* (1883). Kropotkin and Stepniak shared similar political views, and Kropotkin wrote that he felt a real love for Stepniak's honest and frank nature, and his youthful energy and good sense. Stepniak was clearly a man with a keen intelligence and of high moral integrity. The two men were close friends until Stepniak's tragic death from a train accident in London in 1895. Stepniak, in his turn, was to emphasise that Kropotkin was essentially a man of ideas and lacked any real qualities of leadership, 'still less of organising anything'. Kropotkin's strengths, he noted, were as a theoretician and he was almost fanatical in his convictions. Kropotkin was valued in the Chaikovsky Circle primarily as an intelligent and educated person rather than as an instigator of revolutionary action (MK 109-110).

In his much acclaimed narrative of the Russian revolution, Orlando Figes implies (1996: 125-36) that all populists were 'riddled with the guilt of privilege'; made a cult of violence, and romanticised peasant life. Their revolutionary activities are interpreted wholly in negative terms, as expressing a 'totalitarian world-view' or simply as giving the populists a 'sense of belonging'. This negative portrait is completely overdrawn. The notion that populism offered an 'archaic vision of peasant Russia' is almost a caricature, and contrasts with the much more scholarly assessment offered by Venturi and Berlin, who emphasise the progressive and libertarian aspects of the populists.

Figes's book gives the distinct impression that the Tsarist regime was simply 'misguided' or 'mistaken' in its politics (he glides over its totalitarian nature, and the repression, hangings and tortures), while all the revolutionaries were mindless, guilt-ridden terrorists who followed a totalitarian ideology. Neither Herzen, Tolstoy, Chernyshevsky, Lavrov or Kropotkin match this distorted and biased portrait of Russian populism. The Tsarist state was just as much a tragedy for the Russian people as the Bolshevik seizure of power.

What was crucial about the Chaikovsky Circle, as Venturi explores, was that it engendered a communal spirit which gave its members vitality. It evoked a moral atmosphere that was very different from the kind of terrorism expressed by Nechaev. In embracing both the socialist ideas of Lavrov and the anarchism of Bakunin, it expressed a co-operative ethos and a spirit of humanity that was far from being a 'totalitarian' ideology.

As all forms of radical protest in Russia were censored or repressed - even Spinoza's *Ethics* was banned - all meetings and activities of the Chaikovsky circle had to be clandestine. Dressed in workers' clothes, and adopting the pseudonym Borodin, Kropotkin spent two years among the workers of St Petersburg, many of whom were migrant labourers who returned to their peasant villages at harvest time. He spent time distributing radical literature, lecturing, and discussing socialist ideas with them, at meetings or in their homes. Soon the activities of the Circle became known to the secret police - the mysterious and infamous Third Section - which Kropotkin described as a 'state within a state'. For they arrested who they liked, kept people in prison without trial for as long as they pleased, and transported hundreds to Siberia on the mere whim of some general or colonel (MR 336).

By the end of 1873 many of Kropotkin's friends had been arrested, and the Circle became narrower, with meetings more difficult. In March 1874, Kropotkin was himself arrested. The night before, he had given a lecture to the Russian Geographical Society on the origins of the Ice Age, and the glacial formations of Finland and Russia. He was accused of belonging to a secret society, and of conspiracy against the sacred person of the emperor. Kropotkin refused to say anything until brought before a court. He was taken away and imprisoned in the famous fortress of St Peter and St Paul.

A year before he was arrested, Kropotkin had produced what amounted to a revolutionary manifesto of the Chaikovsky Circle, though it expressed essentially his own political views rather than those of the circle. It was entitled: 'Must we occupy ourselves with an examination of the ideal of a future order?' (1873). Though not mentioned in his memoirs, this represents Kropotkin's first major political statement. It is one of the more important documents in the history of the Russian Revolutionary Movement. In advocating the necessity of a social revolution based on agrarian socialism, the manifesto portrays the peasant commune as a *locus* for the transformation of society, not as inherently progressive (MK 101-8). It does not, then, romanticise either the past nor the peasantry. Nor does it advocate terrorism.

5: The Jura Federation and Anarchist Communism

The fortress of St Peter and St Paul has a particular resonance in Russian history, and its very name, Kropotkin recalls, was uttered in St Petersburg in a hushed voice (MR 343). For the tradition of the fortress associates it with suffering, torture and oppression. It was here that the so-called Peter the Great tortured and mutilated the enemies of the imperial state, including his own son. Here the Decembrists were executed in 1825, and many of Kropotkin's contemporaries were imprisoned there: Dostoevsky, Bakunin, Chernyshevsky and Pisarev. It was here too that Karakozov was tortured and hanged in 1866.

On entering this dark, damp and dismal dungeon, Kropotkin immediately thought of Bakunin, who spent six years imprisoned in the Fortress. If Bakunin survived the ordeal, then, Kropotkin said to himself, I must do so too. He was locked in a gloomy cell, which held only an iron bed, and an oak stool and table. A sentry stood outside but would never engage in conversation; Kropotkin found the absolute silence oppressive. His main concern was to preserve his own physical vigour: so he worked out a daily routine of pacing up and down in the cell, aiming to walk daily about five miles.

Kropotkin was at least allowed to read books, and his brother Alexander and his loyal sister Helene had managed to get permission from the Tsar to enable Kropotkin to write, so that he could complete his report to the Geographical Society on the glaciation of Europe. This study was based on his explorations in Finland. He found the work gratifying despite the loneliness and the silence that reigned about him. He engaged in social intercourse, he wrote, only with the pigeons, who came to his window daily to receive scraps of food through the grating. The deadening silence was broken only by the ringing of the bells of the fortress cathedral. As the cell was so damp he began to suffer from rheumatism. Every morning he was taken for half an hour's walk in the prison yard - which he liked.

In spite of the hardships, Kropotkin was therefore cheerful 'continuing to write and draw maps in the darkness, sharpening my lead pencils with a piece of glass which I had managed to get hold of in the yard' (MR 354). But he was very disheartened to hear the news that his brother Alexander had been arrested. A letter Alexander had written to Lavrov had been found in his possession, and he was exiled to Siberia. Kropotkin was never to see his beloved brother again, for Alexander, after spending more than ten years in exile, committed suicide shortly before his expected release in July 1886 (MK 267).

Kropotkin spent two years in the fortress and was visited on one occasion by the Grand Duke Nicholas, brother of the Tsar. The Duke asked lots of questions and tried to engage Kropotkin in conversation - but the anarchist was non-committal. When asked whether he held revolutionary opinions when in the Corps of Pages, Kropotkin replied:

> In the Corps I was a boy, and what is indefinite in boyhood grows definite in manhood (MR 361).

But during his two years in prison, Kropotkin's health steadily deteriorated: he began to suffer from the dreaded scurvy. Many of his comrades in prison had died or committed suicide so that Kropotkin felt increasingly isolated.

While Kropotkin was in prison, the great populist movement 'to the people' began. In the summer months of 1874 (later described as the "mad summer"), thousands of young Russian radicals went as apostles to live among the peasants, wearing peasant dress, doing manual labour in the fields, and trying to enlighten or radicalise the rural population. It ended in disaster. The young radicals met only noncomprehension, suspicion and often downright hostility from the peasants. Socialism, wrote Stepniak to his fellow revolutionary Vera Zasulich in 1876, 'bounced off people like peas from a wall'. The Tsarist state responded to these events with its usual repressive brutality, arresting thousands of radicals, who were then sent without trial, either to prison, or to hard labour in exile (Broido 1978: 85-98, Berlin 1978: 232).

At the end of 1875 Kropotkin was moved from the fortress to a new 'model' prison attached to the St Petersburg court. There was more opportunity for social intercourse with other prisoners, and visitors were permitted more freely. But Kropotkin's health continued to decline. On the advice of a friendly army physician, Kropotkin was therefore transferred, in May 1876, to a small prison attached to the military hospital, situated on the outskirts of St Petersburg. The following month, aided and abetted by a group of friends who devised an intricate and daring plan, Kropotkin made his famous dramatic escape from prison. This escape is graphically described in his memoirs, and makes exciting reading (MR 365-75).

In order to avoid the intensive search being carried out for him, Kropotkin spent some time with a few friends living in a village on the outskirts of St Petersburg. He realised that he had to leave Russia, and after travelling through Finland (in order to avoid the secret agents stationed at the usual border posts) and Norway, he sailed for England. He recorded how satisfying it was to see the Union Jack flying astern the ship, for it signified (in those days) a place of asylum for political refugees.

Kropotkin arrived in Hull in July 1876. He was reluctant to go to London, fearing the presence of Russian spies, and so settled in Edinburgh. He had two overriding objectives. The first was to make contact with James Guillaume, and to indicate his desire to rejoin the Jura Federation. The second was to find some employment. Kropotkin firmly believed (unlike Engels) that a socialist should always rely upon his or her own work to obtain a livelihood. This was difficult, because although he had learned to read and write English, he could hardly speak it. Kropotkin admitted that his Scottish landlady had difficulty in understanding him. However, through his contact with J. Scott Keltie, Kropotkin soon obtained work as a scientific journalist, and began to supply *The Times* and the journal *Nature* with reviews and essays relating to topics of scientific or geographical interest.

Throughout his forty-one years as an exile, such work was to be his primary source of income. But Kropotkin's main aim in life was to further the social revolution that he clearly felt was imminent, and to engage in revolutionary agitation. His inclination, he wrote, drew him more and more intensively towards casting his lot with the working people (MR 379). But he did not stay long in England. Having secured a source of income from scientific journalism, he left for Switzerland in January 1877 to join the Jura Federation. He soon met up with two of his closest friends, Dimitri Klements and Sergei Kravinsky (Stepniak), and settled in La Chaux-de-Fonds.

By that time a real split had occurred in the International Workingmen's Association: between the Marxists, focused in Northern Europe, particularly Germany, and the Bakuninists, whose support came from the Latin countries, particularly Italy and Spain. In Spain more than eighty thousand workers belonged to the International, but it was the Jura Federation in Switzerland that was the heartland of the Bakuninist faction. Kropotkin emphasised that the conflict between the Marxists and the Bakuninists was not of a personal nature:

> It was the necessary conflict between the principles of federalism and those of centralisation; the free

commune and the state's paternal rule; the free action of the masses of the people and the betterment of existing capitalist conditions through legislation - a conflict between the Latin spirit and the German geist (MR 386).

Kropotkin always seemed to identify authoritarian socialism (Marxism) with German culture, and particularly the German imperial state.

The First International more or less came to an end at the conference at The Hague in September 1872 when, to counter the growing influence of the anarchists, the Marxists moved the headquarters of the General Council to New York. But the last conference of the International was actually held at Verviers in Belgium in September 1877, and it seemed when Kropotkin returned to Switzerland that year that even the Jura Federation, which had been the ideological centre of European anarchism, was in decline. But a return to Russia was impossible for Kropotkin. Not only was he too well-known to engage in open propaganda there: the Russian revolutionary movement had become a conspiracy and an armed struggle against autocracy, rather than the popular social movement that Kropotkin envisaged as necessary to invoke a social revolution (MR 378-9).

Between 1877 and 1882, Kropotkin would travel widely through Spain, Germany, France and Belgium, often under the assumed name of Levashov, attending conferences and making contacts with revolutionary socialists. The Jura Federation held a socialist congress each year. The first of these was held at Saint Imier (Switzerland) in August 1877, shortly before the last Congress of the International. The main activities of these congresses were, as Kropotkin put it, 'working out the practical and theoretical aspects of anarchist-socialism' (MR 398). Kropotkin's participation in these various socialist congresses both contributed significantly to the growth of his reputation among European socialists, and provided him with a venue in which to present his own anarchist ideas.

Around the time of the St Imier congress, the Berne authorities banned the carrying of the red flag in the streets of the canton. The socialists considered this an infringement of their rights as citizens. They therefore decided to defy the ruling, and on the anniversary of the Paris Commune, they marched through Berne carrying a red flag. Although many of the demonstrators were armed, only minor injuries were incurred in the clash with the police, and there was no serious bloodshed (MR 397-98). As Woodcock and Avakumovic write:

> This was the nearest Kropotkin ever came to revolutionary street-fighting, and he certainly seems to have shown a spirit rather different from that of the gentle scholar (WA 161).

At the Jura Federation's congress held in Fribourg (Switzerland) in August 1878, attended by only eight delegates - Guillaume was conspicuously absent - Kropotkin presented his first major political statement. In his address to the delegates he advocated the negation of the nation-state, which he considered then to be in the process of disintegration, and its replacement by a free federation of communes and productive associations. This implied also a social revolution, a fundamental social change, during which, through 'insurrectionary deeds', people would spontaneously expropriate land, capital and the means of livelihood.

In the following year, at the Federation's congress held at La Chaux-de-Fonds in October 1879, Kropotkin gave a similar address, entitled 'The Anarchist Idea from the Viewpoint of its Practical Realisation'. Martin Miller indicates three important points about this speech: that Kropotkin refused to advocate the formation of the anarchists into a political party and thus seek power through the state; that the coming social revolution was defined in terms of the collective expropriation of land, capital and the means of production; and, finally, that Kropotkin still continued to use the Bakuninist term 'collectivism' (MK 142).

At the Jura Federation congress of October 1880, also held at La Chaux-de-Fonds, Kropotkin abandoned the concept of 'collectivism', which he felt still implied a system of wage-labour, and advocated for the first time 'anarchist communism'. This entailed the free distribution of goods: the notion of an economic system based on the adage 'from each according to his means, to each according to his needs'. The congress thus affirmed the formation of anarchist communism as a distinctive form of socialism. For Kropotkin this was an 'important stage' in the development of revolutionary socialism. Many of the delegates at the congress, however, were hesitant about the term 'communism', for in France it was intrinsically associated with monastic life.

Although Kropotkin played an important part in the development of anarchist communism, and was later to become its chief exponent and advocate, he was not its originator. Already in 1876 François Dumertheray, who was later to help Kropotkin to establish the newspaper *La Révolte*, had published a little pamphlet in Geneva entitled 'Aux Travailleurs Manuels Partisans de l'Action Politique', which had advocated anarchist communism. The linkage between anarchism and communism indeed seems to have evolved spontaneously and independently among many of the 'collectivist' followers of Bakunin in Italy, Spain and Switzerland. People important in the development of anarchist communism, besides Kropotkin, include Elisée Reclus, Carlo Cafiero, Jean Grave and Errico Malatesta (Cahm 1989: 51-64).

It was through the Jura Federation congresses that Kropotkin met many anarchist comrades who were later to become famous revolutionaries, many also becoming firm and life-long friends. These include: James Guillaume, Elisée Reclus, Errico Malatesta, Adhemar Schwitzguebel, Paul Brousse, Carlo Cafiero and Nicholas Zhukovsky (MR 391-95). Of particular importance was Elisée Reclus, who, like Gustave Lefrançois and Louise Michel, was an ex-Communard. Like Kropotkin, Reclus became a celebrated geographer. A close friend of Bakunin, Reclus had been an anarchist for several years before he met Kropotkin. At their first meeting they appear to have had a serious argument, but later they became intimate friends and associates. Kropotkin considered Reclus a "true puritan" in his manner - Reclus was a vegetarian - but one who had the mind of a French encylopaedist philosopher. Kropotkin considered Reclus's books to be among the very best of the century (MR 392). Reclus, in turn, always paid tribute to Kropotkin's intellect and moral integrity. In the preface to his *Paroles d'un Révolté* (*Words of a Rebel*, a collection of Kropotkin's writings), published while Kropotkin was undergoing a second period of imprisonment (1885), Reclus would write of his friend:

> Among those who have observed his life from near and far, there is nobody who does not respect him, who does not bear witness to his great intelligence and his heart overflowing with goodwill; there is nobody who will not acknowledge his nobility and purity of nature.

Kropotkin's only crime, Reclus concluded, was to love and defend the cause of the poor and powerless (WR 16).

At the end of 1878 Kropotkin met and married a young Polish Jew, Sophie Ananiev, then aged twenty-five. Born in Kiev, she had come to Switzerland to study, and had secured a place studying biology at Berne University. In a letter to his friend Paul Robin, Kropotkin describes her as quiet, kind and gentle, and as having a 'wonderful disposition'. In his memoirs Kropotkin does not mention his marriage, and hardly refers to his wife Sophie at all in the text. Their marriage, however, seems to have been a happy one, and although Sophie has been described as a 'dedicated and subordinate companion' of Kropotkin, she was clearly an intelligent and independent woman. Kropotkin notes that he always discussed all events and papers with Sophie, and that she was a 'severe literary critic of my writings' (MK 297, MR 424). Though strong-willed and intelligent, she did not match Kropotkin's intellectual stature, and her attitude towards him, from the moment they met, was one of devotion and admiration. Her one fault was a lingering snobbery, as she like to be recognised as a 'princess' (WA 172). Kropotkin was equally devoted to Sophie. They had one child, Alexandra, who was born in London in April 1887.

Around 1879 the outlook for the Jura Federation seemed bleak. The last issue of the Bulletin of the Federation had been printed in 1877, its publication having been discontinued through lack of subscriptions. Guillaume had temporarily retired from active politics. In Geneva Kropotkin had collaborated with Paul Brousse in editing a small paper *L'Avant Garde*. But the paper had published articles commending the terrorist attacks on European rulers. In consequence Brousse had been imprisoned, and the paper - the only remaining anarchist publication - suppressed by the Swiss authorities.

It was in these very disheartening circumstances that Kropotkin decided to launch the publication of a new anarchist journal. As none of the leading anarchists for various personal reasons, were willing or able to accept responsibility for its publication, Kropotkin eventually obtained the support and help of two Geneva working men, George Herzig and François Dumartheray. They only had about twenty-three francs (then about four dollars), between them to start the new fortnightly paper but they decided to be bold and to print two thousand copies. (Until then no local anarchist paper had sold more than six hundred copies!) The new paper, *La Révolte,* was launched in February 1879. Kropotkin had to do most of the writing himself. It was, he wrote, 'moderate in tone, but revolutionary in substance, and I did my best to write it in such a style that complex historical and economic questions should be comprehensible to every intelligent worker' (MR 418).

The paper proved a great success. Unfortunately, the printer was threatened with losing his lucrative government printing contracts should he continue to publish the subversive *La Révolte.* He was thus forced to withdraw as publisher. Kropotkin and his two Swiss friends therefore set up their own printing press: Imprimerie Jurasienne. In his memoirs Kropotkin speaks highly of Herzig and Dumartheray's contribution to the publications as well as of the support he received from Elisée Reclus (MR 417-23). As noted above, Reclus collected the essays that Kropotkin had written for *La Révolte* and published them in Paris in 1885 as *Paroles d'un Révolté - Words of a Rebel.*

Kropotkin's wife Sophie was suffering from the cold winds of Geneva, so on the advice of her doctor, Kropotkin moved to Clarens in the spring of 1880. Here Elisée Reclus lived at that time. The couple settled above Clarens in a small cottage overlooking the blue waters of Lake Geneva and with the snow-capped mountains behind. Clarens was near enough to Geneva for Kropotkin to retain his contacts with his anarchist friends, and it was here that Kropotkin wrote some of his best articles for *La Révolte*, and laid the foundations for all his subsequent writings. Kropotkin was also invited by Reclus to contribute to his *Geographie Universelle* work which he combined with the writing of anarchist propaganda (MR 424).

While Kropotkin was living in Switzerland the political situation in Russia had become increasingly acute, with all forms of radical protest being repressed with brutal and senseless ferocity. Hard labour from six to twelve years in the mines and subsequent exile to Siberia for life was a common sentence. Kropotkin notes that one young woman got nine years' hard labour and life exile to Siberia merely for giving a socialist pamphlet to a worker. The Tsarist state, he wrote, was in a state of siege, and hanging in Russia had become the order of the day (MR 425-7).

The culmination of this state of affairs was the assassination of Alexander II in March 1881. This caused an extreme reaction throughout Europe. In Russia, a secret reactionary society, the Holy League, was formed under the leadership of the Tsar's brother. Largely recruited from the officer class, its function was to combat revolutionary activity. There were rumours that Kropotkin's own life was threatened by the League. Of more practical import, Kropotkin was expelled from Switzerland in August 1881. This was because articles in *La Révolte* had supported the actions of the People's Will (*Narodnaya Volya*), the group that had carried out the execution of Alexander II, although Kropotkin himself was never an advocate of regicide or individual terrorism.

A month earlier, in July 1881, Kropotkin had attended the International Anarchist Congress in London, along with about forty-five other delegates from a variety of countries and organisations. Here Kropotkin discussed political issues with the likes of Nicholas Chaikovsky, Louise Michel, Errico Malatesta, Emile Gautier, and the British socialists Frank Kitz and Joseph Lane (WA 179).

After his expulsion from Switzerland in August 1881, Kropotkin and his wife Sophie settled at Thonon, just over the French border from Geneva, to enable Sophie to complete her bachelor of science degree at the University of Geneva. Around October, they moved to London and went to live in the 'seedy district' of Islington. Here Kropotkin renewed his acquaintance with the radical liberal MP Joseph Cohen, as well as meeting the Marxist, Henry Hyndman, who that same year (1881) founded the Social Democratic Federation. Hyndman was later to record that Kropotkin at that time was overflowing with enthusiasm and vigour.

> At first I tried to argue with him about his anarchist opinions, which seemed to me entirely out of accord with his intelligence and naturally charming disposition. I found this was quite hopeless (WA 185).

In spite of their political differences, the two socialists always remained friends. Henry Hyndman introduced Kropotkin to James Knowles, the admirable and enterprising editor of the journal *Nineteenth Century*, and Knowles commissioned Kropotkin to write a series of articles on Russian prisons.

This was the beginning of an association and a friendship that was to last thirty years. Through Joseph Cohen, Kropotkin also wrote several articles on Russian affairs for the *Newcastle Chronicle*. But though active, Kropotkin was unhappy in London. In his memoirs he describes it as a 'year of real exile': for someone who was a revolutionary socialist 'there was no atmosphere to breathe in'. With his friend Chaikovsky he began to spread socialist propaganda amongst London workers. He also went to speak at radical clubs about Russian affairs, but seldom more than a dozen people ever turned up. Kropotkin records that both he and his wife felt lonely in London, and when they left in October 1882 they often said to each other 'Better a French prison than this grave' (MR 440-42).

They returned to the French town of Thonon, but only enjoyed a brief two months of freedom, for in December 1882 Kropotkin was arrested and taken to Lyons. His wife's brother had died in his arms only the night before, but the police refused to allow Kropotkin to stay and comfort Sophie.

In January 1883 Kropotkin's trial was held in Lyons. He was accused, along with sixty-five other anarchists, of belonging to an illegal organisation, the International. The reason for these arrests is that there had been an upsurge of strikes and insurrections among the workers of Lyons, but Kropotkin himself had in no way been directly involved in these incidents. The International Workingmen's Association, in any case, had long ceased to exist as a viable organisation. Nonetheless, Kropotkin was found guilty and condemned to five years in prison. He made a propaganda speech in court, and signed a statement of principles which he and his co-defenders had drafted. It was a manifesto that gave a succinct outline of anarchist principles (WA 190).

After a couple of months in the Lyons prison, Kropotkin was moved, in March 1883, along with twenty-two other anarchist prisoners, in great secrecy to the central prison at Clairvaux. The prison was formerly the Abbey of St Bernard. Conditions in this prison were not so harsh as in Russia, and Kropotkin was able to continue his literacy work, writing articles for the *Encyclopaedia Britannica* and *Nineteenth Century*. The prisoners were able to organise educational classes amongst themselves, and Kropotkin even cultivated a small garden in which he grew lettuce and other vegetables. His wife Sophie abandoned her studies in Paris, and came to live in the tiny hamlet of Clairvaux so that she could be near Kropotkin, even though during the first year she was only allowed to visit him infrequently.

While in prison Kropotkin also began to study the detrimental effects of prisons on the human personality, and the unjust and unhealthy nature of these institutions. Kropotkin's observations and reflections of prison life were set forth in his book *In Russian and French Prisons* published in London in 1887. The book had a curious history, in that it immediately disappeared from the bookshelves, the whole edition having been bought-up and destroyed by an agent of the Russian government. Such is the power of the written word! It was later reissued by another publisher.

Kropotkin's second spell in prison lasted three years, and towards the end he began to suffer from malaria and scurvy, as Clairvaux prison was built over marshy ground. Appeals on Kropotkin's behalf were organised in both Paris and London by a variety of friends, scholars, writers and revolutionaries. Petitions for his early release were compiled, and the signatories included such well-known figures as Leslie Stephen, Patrick Geddes, Alfred Russel Wallace, William Morris and H. W. Bates. But the French government, cognisant of their alliance with Imperial Russia, kept stalling, hinting that 'diplomatic difficulties' stood in the way of Kropotkin's release.

Eventually, under pressure, the French government succumbed, sensing that Kropotkin would be much less trouble if he were out of the country. Kropotkin was finally released from Clairvaux prison in January 1886. Having escaped from prison in Russia, been expelled by the Swiss authorities from Switzerland, and now banned from France, where he never felt an exile, Kropotkin and his wife headed for England. After spending a few weeks in Paris, visiting Kropotkin's friends Jean Grave and the Reclus brothers, they arrived in London in March.

6: The Socialist Movement in England

On their arrival in London, the Kropotkins lived at first with Sergei Kravchinsky (Stepniak). They stayed with him about a month in St John's Wood until they had established their own house - a modest cottage in Harrow, where Kropotkin's other old friend Nicholas Chaikovsky also came to live. After having spent three years in prison, Kropotkin felt at this time physically and psychologically exhausted, and had to refuse an offer from William Morris to write articles for his journal *Commonweal*. But at the end of the year further tragedy struck. First, his wife Sophie fell seriously ill with a bout of typhus, and then, in the autumn, he learned of his brother's suicide while in exile in Siberia. 'A dark cloud hung over the cottage for many months' he wrote in his memoirs (MR 491).

But during the four years that Kropotkin had been away from London, the political situation in the country had completely transformed, and, as Kropotkin wrote, by 1886 the socialist movement in England was 'in full swing'. This socialist movement, broadly speaking, consisted of three major groupings.

First, there was the Social Democratic Federation - a development of the radical Democratic Federation founded in 1881 by Henry Hyndman. In 1884 it had become socialist, and had added the term 'Social'. Hyndman was greatly influenced by Karl Marx, even though Marx himself considered Hyndman's little book *England for All* largely a work of plagiarism - of his own treatise on capital, *Das Kapital*. Hyndman, an ex-stockbroker, has been described as a 'Tory Democrat', for he expressed jingoistic and imperialist sentiments, but the policy of the SDF was consistently Marxist.

In 1884 there had been a split in the Social Democratic Federation. It focused around Andreas Scheu, a Russian refugee and an anarchist associate of Johann Most, and William Morris. They both resented the domineering attitude and political opportunism of Hyndman, and advocated an anti-parliamentary socialism. This led to the formation of the Socialist League, which included among its members Frank Kitz, Charles Mowbray, and Stepniak. Its manifesto, drafted by Morris, John Quail describes as a 'beautiful' document, which, if not anarchist, 'is clearly libertarian in its commitment to revolution, its view of the role of socialist groups and its depreciation of state and party hierarchy' (1978: 23-38).

The third socialist grouping was the Fabian Society, also founded in 1884. Although begun under the influence of the Democratic Federation, it was much more middle-class and, from its inception, reformist in its politics. Associated as it developed with the likes of George Bernard Shaw, Sydney and Beatrice Webb, and Graham Wallas, it always championed a form of parliamentary state socialism. The Fabians were later instrumental in the foundation, around the turn of the century, of the British Labour Party.

Although at that period there was no organised anarchist movement in Britain, there were a number of individuals who expressed anarchist tendencies. Both Frank Kitz and Joseph Lane had attended the International Anarchist Congress in London in 1881, and Lane (1851-1920) had been involved in radical activities since he was a boy, when he had rebelled against the infamous game and land laws. From a rural working-class background - he was born near Wallingford in Oxfordshire - Lane throughout his early life had been involved in various working-class organisations in London, including the Manhood Suffrage League and the Honerton Section of the Social Democratic Club. He recalled having met Kropotkin and Malatesta at the 1881 congress.

Having formed the Labour Emancipation League the following year (along with Kitz and others), Lane was one of those who left the SDF to join the Socialist League, and he became manager of its journal, *The Commonweal*. Although primarily an activist, in 1887 Lane published his famous pamphlet *An Anti-Statist Communist Manifesto*. This was the first English anarchist pamphlet for more than half a century, and in it Lane proposed a form of revolutionary socialism that was close to Kropotkin's anarchist communism. Advocating atheism, anti-statism and communism, Lane wrote that 'the object of socialism is to constitute a society founded on labour and science, on liberty, equality and solidarity of all human beings' (Lane 1978: 27).

Frank Kitz (1849-1923) was an associate of Lane, and formed an important link between the foreign emigrés to London like Malatesta and Most - who were anarchists - and the new socialist organisations that were emerging in the 1880s. He, too, was of working-class origins, and was involved, like Lane, in many working-class organisations and clubs, specifically those devoted to free thought and the revival of socialism. Described as a 'colourful character', a rebel in temperament rather than an anarchist by philosophy, Kitz was an active propagandist for revolutionary socialism. In his published recollections (1912) he makes no mention of Kropotkin, but is critical of both Hyndman and his 'jingoism' or Tory socialism, and of the Fabian Society and its 'suburban socialism'. Describing himself as a 'wanderer', with an intense love of the countryside, he closely identified with William Morris who, he writes, saw the debasement of art and the destruction of natural beauty as an inevitable outcome of capitalism (1912: 21).

But anarchist ideas were also being expressed in more middle-class circles. Two figures are of particular significance in the emergence of anarchism in Britain. The first of these is Henry

Seymour (1862-1938). Seymour was a radical secularist and freethinker from Tunbridge Wells who achieved early notoriety by being prosecuted for blasphemy. Around 1884 he converted to anarchism, influenced by reading the works of Spencer, Bakunin and Tucker. He came to advocate an individualist form of anarchism in the Proudhonist tradition with an emphasis on individual property. In March 1885 he established the first English-language anarchist journal *The Anarchist*, joining the Fabian Society in that same year. At meetings of the Society he met Charlotte Wilson (1854-1944), a well-educated, middle-class woman who was married to a stockbroker. She had been a founder-member of the Fabian Society, and was the only woman elected to the first executive committee of the Society, although she is hardly mentioned in Margaret Coles's history of Fabian Socialism (1961).

Towards the end of 1884 Charlotte Wilson too converted to anarchism, and contributed to a series of articles on 'Anarchism' to *Justice*, the paper of the Social Democratic Federation. Two years later she wrote a short piece on 'Anarchism' in the Fabian Tract (no 4) on "What Socialism is" (June 1886). What seems to have particularly influenced Wilson's shift towards anarchism was reading in *The Times* an account of Kropotkin's trial in January 1883. She afterwards wrote that the 'noble words' of Kropotkin echoed in the hearts of all seekers after truth - herself among them. She thus became a firm devotee of Kropotkin's ideas on anarchist communism, and seems to have written to Sophie Kropotkin in Clairvaux (Oliver 1983: 28). But although Wilson was perhaps the best known English anarchist in the 1880s, she does not seem to have had much contact with working-class activists in the trade unions or socialist organisations (Walter, in Wilson 1979). Although later a radical suffragette, Wilson eventually dropped out of the anarchist movement, but at the time she met and collaborated with Kropotkin, she was a dedicated and enthusiastic advocate of anarchist communism.

Around March 1886 Kropotkin joined a small circle of anarchists which became known as the London Anarchist Group of Freedom. Its aim was to distribute anarchist propaganda. Wilson was the leading member of this group, and was its representative at the Fabian congress in June 1886. The group made an effort to co-operate with Henry Seymour, the editor of *The Anarchist*, but did not succeed, although Seymour was quite happy to publish material from all anarchist traditions and did, in fact, publish extracts from the writings of Elisée Reclus and Kropotkin.

There was, however, evidently a good deal of personal friction between Seymour and Wilson, while Kropotkin himself was anxious to develop a journal that was exclusively devoted to the kind of anarchist communism that had been formulated in the Jura Federation. Stepniak, who was a close friend of Kropotkin, and himself a dedicated revolutionary socialist, stressed in his memoir Kropotkin's single-mindedness. Although Stepniak - Sergei Kravchinsky - more than anyone respected Kropotkin's theoretical brilliance and his admirable personal qualities, he also wrote that Kropotkin 'is too exclusive and rigid in his theoretical convictions', in that he admits no departure from his own ultra-anarchist programme. Kropotkin was clearly at odds with the individualist anarchism advocated by Proudhon, Tucker and Seymour. But although the picture of Kropotkin as the 'gentle anarchist prince' may be somewhat overdrawn, to describe Kropotkin as an 'autocrat' is, I think, equally misleading (Quail 1978: 52).

Kropotkin and his friends therefore decided to disassociate themselves from Seymour and *The Anarchist* and in October 1886 the first issue of *Freedom* appeared, and Charlotte Wilson's group became the Freedom Press. The first issue of *Freedom* was a four-page sheet, mostly written by Kropotkin. Wilson was the editor and publisher of *Freedom* and its main supporter from 1886 to 1894, when for domestic reasons she left the movement.

Freedom and the Freedom Press (which is now one of the main publishers of anarchist literature in Britain) are still flourishing after more than a hundred years. Kropotkin himself was to remain in close touch with *Freedom* for the next three decades (until 1914) and many of his early seminal articles on anarchist communism were published in *Freedom* between 1886 to 1890. They were collected and published as *Act for yourselves* (1988).

After his wife's illness and the tragic news of his brother's suicide later in 1886, Kropotkin threw himself wholeheartedly into anarchist propaganda, and throughout the winter months he lectured all over the country, visiting all the important industrial centres of England and Wales. He met people of all classes, and participated in socialist gatherings in Cambridge, Edinburgh and Glasgow, speaking to enthusiastic crowds. In Newcastle in October some four thousand people listened to his address on the modern development of socialism. Kropotkin was clearly excited with this development, writing to Peter Lavrov that 'socialism has become **the** question of the day' (MK 166). But Kropotkin later reflected that, while in Switzerland he had hardly met during his four years there any but working class people, all his anarchist associates in England were middle-class (MR 495). When, more than a decade later, revolutionary socialism had gone into decline, and workers had become increasingly involved in party politics, Kropotkin was to describe British anarchism, in a letter to his friend Herzig (March 1904), as

> Anarchie de Salon - epicurean, a little Nietzschean, very snobbish, very proper, a little too Christian (MK 169).

Contemporary life-style or post-modern anarchism is hardly a new phenomenon.

The respected historian of anarchism Max Nettlau (1865-1944), who had been a member of the Socialist League (which he described as the flower of English revolutionary socialism) always regretted the fact that Kropotkin never joined the League. 'I have always felt', he wrote, 'that a splendid opportunity was lost here' (1992 (1921): 384).

In 1887 Kropotkin resumed his scientific journalism in earnest, and began to contribute notes and articles for *Nature* and *The Times*. This was to be his only source of income. Kropotkin never borrowed or accepted any gifts of money, nor would he take payment for the work he did on behalf of anarchism. All he needed was earned by his writing, and Kropotkin always appears to have lived fairly modestly (WA 211-13). He also became involved that same year in demonstrations relating to the Chicago anarchists who had been falsely accused of throwing the bomb that killed some policeman at a political rally, held to agitate for an eight hour day. The men were hanged in November 1887, and two days later Kropotkin took part in a large gathering in Trafalgar Square, demonstrating against existing working conditions. Scuffles broke out with the police, and there were several arrests. Woodcock and Avakumovic write that 'Kropotkin's part in this incident was slight, but that he should have gone at all, in his insecure position as an alien in the one country that offered a free refuge, showed the depth of his concern for such concrete issues' (WA 215).

In 1887 Kropotkin also wrote two important articles on anarchism for the magazine *Nineteenth Century*. These were afterwards published by the Freedom Press as the pamphlet *Anarchist Communism: its Basis and Principles* (1891).

During this period Kropotkin became a close friend of William Morris (1834-1896). The Socialist League was then experiencing an ideological schism between the parliamentarians, led by Belfort Bax and Eleanor Marx, and the libertarian

Socialists. Joseph Lane's *Manifesto* was an expression of the anti-statist faction. There was a close affinity between the libertarian socialism of Morris and Kropotkin's own anarchist communism, but Morris would never come to call himself an anarchist. He seems to have seen an absolute dichotomy between anarchism and communism, and tended to equate the former with individualist anarchism (Morton 1973: 209-212). Neither Kitz nor Lane described themselves as anarchists, tending rather to see themselves as revolutionary socialists - but all three men, Lane, Kitz and William Morris (the anarchist wing of the Socialist League) expressed political ideas and principles that were almost identical to those of Kropotkin.

The decade of the 1890s has been seen as a time when Kropotkin took on the persona of a 'saintly scholar' for during this period he returned to his scientific work; made contact with a host of important scholars and institutions, and began writing a series of essays and reviews that when published in book form were to make him famous. All these books are written in a lucid and engaging style, and have a simplicity and brevity of expression that is unusual in political and scientific texts. His books are inspiring, and refreshingly free of the obscurantist jargon that these days sometimes passes for scholarship in the groves of academia, so they reach out to a wide audience. Kropotkin, like the naturalist Jean-Henri Fabre, believed that ideas, even though complex, should always be made understandable to the common person. (The term "common" for Kropotkin carries no negative connotation).

These books include: *The Conquest of Bread* (1892), originally written as a series of articles for *La Révolte* between 1886-91, when Kropotkin's friend Jean Grave took over its editorship; *Fields, Factories and Workshops* (1899) based on a series of essays published in *Nineteenth Century* between 1888-1890; his *Memoirs of a Revolutionist* (1899), first issued in *Atlantic Monthly* in 1898-99, and *Mutual Aid* (1902), also first published as a series of essays in *Nineteenth Century* between 1890 and 1896.

So by the 1890s Kropotkin had settled down in England, his last place of refuge. He established a pattern of life that combined manual labour (for Kropotkin and his wife were both enthusiastic and serious gardeners), political agitation, and the retired life of a scholar. His political activities however were largely confined to propaganda: 'to lecturing, writing, and occasional attendance at reunions and celebrations organised by anarchist groups and London working men's clubs' (WA 219).

Kropotkin established close friendships or intellectual contacts with many important contemporary figures. These included the famous naturalist Henry W. Bates, who wrote *A Naturalist on the Amazons*, and who was Secretary of the Royal Geographical Society; W. Robertson Smith, Professor of Arabic at Cambridge University and a pioneer of anthropology; Keir Hardie of the Independent Labour Party; the much neglected sociologist Patrick Geddes, as well as libertarian socialists like Edward Carpenter and William Morris.

What is quite remarkable and consistent is that almost all of his contemporaries describe Kropotkin - his intellectual stature, demeanour and personality - in the most glowing terms. At this time Kropotkin is described as a short, burly man with a large, unrestrained beard and a bald head, and as having sparkling, almost incandescent eyes. He is described as kind, hospitable, modest, warm and affectionate and as 'brimming over with life and interest in everything'. His outstanding intellectual abilities, his fluency in several languages, and his moral integrity were always affirmed by his contemporaries. Oscar Wilde, himself a libertarian socialist and an acute observer of human life, described Kropotkin as being almost Christ-like, and as living one of the most perfect lives he had ever encountered (*De Profundis* 1905: 75). George Bernard Shaw likewise recorded that Kropotkin 'was amiable to the point of saintliness, and with his red full beard and loveable expression might have been a shepherd from the 'Delectable Mountains' (WA 225). Romain Rolland, who was a great admirer of Tolstoy (and Gandhi) said of Kropotkin:

> Simply, naturally, has he realised in his own life the ideal of moral purity, of severe abnegation, of perfect love of humanity that the tormented genius of Tolstoy desired all his life (WA 267).

But Kropotkin was a human being like the rest of us; he had his faults and blind spots, he made mistakes, he had his prejudices and inconstancies, but throughout his long life he held fast to the commitments that he had made early in life, and he thus remained a revolutionary socialist.

During the 1890s Kropotkin began to travel extensively. On his lecture tours in Britain he never lost the opportunity to visit factories, coal mines and workshops. Also, unlike most socialists and ecologists, Kropotkin was intensively interested in agriculture and in the production of food: while in prison in Clairvaux and during all his years in exile, he always maintained a kitchen garden. He made several visits to the Channel Islands in order to study market gardens. Many of the ideas expressed in *Field, Factories and Workshops* are based on first-hand experience, and indicate an intimate knowledge of horticulture and agricultural practices (WA 237).

In 1892 the Kropotkins left Harrow. After living for a while at Acton, they settled, in the late summer of 1894, in Bromley, Kent, where they acquired a small cottage. The Kropotkin's home soon became an 'open house' for a host of visitors - anarchist exiles like Errico Malatesta, Louise Michel, Rudolf Rocker, Alexander Atabekian and Varlaam Cherkesov, as well as radical scholars and British socialists such as Keir Hardie, Ben Tillett, John Burns and Tom Mann.

But in the years after the London dock strike of 1889, events took a sharp turn, for while Kropotkin became increasingly respectable as an anarchist 'savant', anarchism as a movement in England began to decline. Two reasons have been suggested for this decline. The first was that the British Labour movement turned increasingly away from revolutionary socialism, and began to support 'law and order' and 'moderation'. This led, Kropotkin argued, to the decay of the whole socialist movement. Led by Hardie, Tillett and Burns, the labour movement went 'deeper and deeper into the quagmire of parliamentary politics' (AY 120). Only Morris and Kropotkin seem to have opposed this trend.

In 1893 the Independent Labour Party was formed. Led by Keir Hardie, and with Tom Mann as its secretary, it joined forces with the Social Democratic Federation and the Fabians in advocating parliamentary action. In 1893 the Zurich congress of the Second International, dominated by Marxists, voted that the anti-parliamentarian socialists should be excluded from the International. Eventually Hardie, Tillett and Burns would become MPs, with Burns a cabinet minister in the Liberal government, although he remained true to his radicalism in opposing Britain's entry into the First World War (Cole 1954: 413). Tom Mann (1856-1941) later became a revolutionary syndicalist.

The second reason for the decline of anarchism as a movement in the last decades of the century, was that terrorism became increasingly identified with anarchism. Terrorism, or what became known as 'propaganda by deed' had always been an intrinsic part of the radical movement in Russia, in its struggles against Tsarist autocracy and oppression. But in the 1890s isolated terrorist acts began to erupt throughout many parts of Europe: for example, in December 1893 Auguste Vaillant threw a home-made bomb into the French Chamber of Deputies (it killed no one), while in June 1894 President Carnot was assassinated at Lyons by the Italian anarchist Santo Caserio. Kropotkin never advocated individual acts of terrorism: what he did advocate was a social revolution, the transformation of property relations by a popular act of expropriation, as when the peasants, during the French revolution, took possession of the land. But unlike the bourgeoisie and the Tolstoyan pacifist anarchists who completely repudiated and decried such terrorist acts, Kropotkin, like other members of the Freedom Group, pointed out that the men who committed these acts 'were often placed in intolerable circumstances by the injustice they had to see and endure'. Kropotkin, however, left no doubt that he did not himself advocate such methods (WA 243).

Unfortunately, the violent acts of individual anarchists resulted in severe repressive measures being taken by governments throughout Europe. In France, the police suppressed *La Révolte*, the chief anarchist newspaper founded by Kropotkin, and in March 1894 its editor Jean Grave was imprisoned. A year later, when he was released, Grave, with the encouragement of Reclus and Kropotkin, founded the paper *Les Temps Nouveaux*. The first issue was published in May 1895, and it flourished as an anarchist paper until the outbreak of the First World War. Kropotkin was to contribute many articles to it.

Early in 1896, Grave decided to hold a series of lectures in Paris in order to raise funds for the new paper. Kropotkin was invited to give a lecture, and an audience of five thousand people was expected. However the lecture was never delivered, as the French authorities at Dieppe, wary of Kropotkin's influence refused him permission to enter the country. Although Kropotkin has often been described as an inactive, scholarly recluse during his exile in Britain, doing nothing but 'talking and writing', such talking and writing, as Roderick Kedward affirms, had the effect of being 'potent agitation' and was greatly feared by the French government (1971: 53-54). The lecture Kropotkin was to have given was later published as 'The State: Its Historic Role' (1896). It is one of the best-known of Kropotkin's political essays.

In 1896 Kropotkin was saddened by the death of two close friends: William Morris, who died in the October, and his comrade of the Chaikovsky Circle, Stepniak, who was tragically killed at a railway crossing in Surrey two months later. Kropotkin treasured the memory of both these gifted men. The following year, Kropotkin made his first visit to the United States and Canada, visiting his friend James Mavor, who was Professor of Economics at Toronto University. Through Mavor, Kropotkin influenced the naturalist Ernest Thompson Seton, who in 1902 was to found the first outdoor youth movement: the Woodcraft Indians.

Kropotkin gave lectures on geomorphology at the University, and travelled across Canada, visiting all principle towns. He became particularly interested in the Mennonite settlements in Canada, which were based on communal agriculture and sound farming practices. He also made a brief tour of the United States, lecturing in cities throughout the Eastern Seaboard: Chicago, Boston, Philadelphia, New York. Around two thousand people attended many of these lectures, and they were well received (Avrich 1988: 84). Kropotkin refused to meet Andrew Carnegie but went out of his way to make contact with many anarchists, including Johann Most, Harry Kelly, Emma Goldman, Volterine de Cleyre and Benjamin Tucker. He travelled to Pittsburg hoping to see Alexander Berkman, but was not allowed to meet him as Berkman was in solitary confinement.

Three important things came out of Kropotkin's North-American tour of 1897. The first was an invitation from the editor of *Atlantic Monthly* to write his memoirs. The outcome was an autobiography of extraordinary richness, one that largely depicts Kropotkin's early life as a nineteenth-century Russian revolutionary. His book remains a classic in this literary genre, and his own story is told with modesty and with poignancy (WA 282).

The second outcome was that Kropotkin, through Mavor, was to be instrumental in finding a safe haven for the Doukhobor religious community, who had long been harassed by the Tsarist regime for their pacifism. Supported by Tolstoy, nearly twenty thousand Doukhobors eventually left Russia to settle in Western Canada (WA 283).

Thirdly, Kropotkin's visit injected a new lease of life into the anarchist movement in the United States. This is affirmed by Emma Goldman, who described Kropotkin as 'our beloved teacher'. Although, compared with Bakunin and Most, Kropotkin was not a dynamic speaker, his lectures were given with enthusiasm and emotion, and he always made a deep and lasting impression on his audience. Anarchists were described as being 'jubilant' over Kropotkin's lectures, for they had given the movement a badly needed lift (Avrich 1988: 86-87).

7: The Coming Revolution

Kropotkin had never completely lost contact with his native Russia, although many of the Russian emigrés he knew in London, apart from Cherkezov, tended to be constitutional liberals. Valaam Cherkezov came from Georgia, and, like, Kropotkin, from the highest echelons of the Russian aristocracy. He was Kropotkin's best-known associate in London, and was a strident critic of Marxism. It has been suggested that Cherkezov was the only Russian with whom Kropotkin collaborated during his early years of exile in London.

Ever since he was in Switzerland around 1879, Kropotkin had despaired of the populist movement in Russia, which had increasingly emphasised constitutional reform and had adopted conspiratorial methods of terrorism. A small group of Russian exiles, however, based in Geneva - of whom the Armenian Alexander Atabekian, a devotee of Kropotkin, was the most prominent – had established in 1892 an anarchist propaganda circle. The group called itself the 'Anarchist Library' and began printing anarchist pamphlets. The first of these was Bakunin's *The Paris Commune and the Nature of the State*. It had an introduction by Kropotkin. This Genevan group of Russian exiles also began publishing, in pamphlet form, the works of other Western European anarchists. Besides Kropotkin, these included Jean Grave, Elisée Reclus, Errico Malatesta and Saverio Merlino (Avrich 1967: 38). The group also began to make contact with individuals in Russia.

But Kropotkin's attention and interest in Russia was further revived in 1901 when on a return visit to the United States, he was invited to give an important series of lectures at the Lowell Institute in Boston, on Russian literature. They were later published in book form as *Russian Literature: Ideals and*

Realities (1905). As open political discussion was prohibited in Tsarist Russia - indeed **all** political activity was banned - the only way people had of expressing their discontents and aspirations was through literature: novels, poems, satires and literary criticism. Indeed, Kropotkin was to write that 'In no other country does literature occupy so influential a position as it does in Russia. Nowhere else does it exercise so profound and so direct an influence upon the intellectual development of the younger generation' (RL xi).

Kropotkin's book was intended only to give a broad, general idea of Russian literature, and at this level it succeeds, providing a good and comprehensive introduction to this literature up to the end of the nineteenth century. It is thus more a book of literary history and appreciation - Kropotkin is warmly sympathetic towards all well-known Russian authors (Pushkin, Gogol, Turgenev, Tolstoy, Dostoevsky) - than a work of literary criticism. As George Woodcock writes, it hardly counts as criticism of any profound and penetrating kind, and given Kropotkin's own serene temperament and political views, he found it hard to understand the spirituality and the psychological angst that motivated the great novels of Tolstoy and Dostoevsky.

Kropotkin felt much closer to Turgenev whose novels, though penetrating, are devoid of the moralism of Tolstoy or the morbidity and torment of Dostoevsky (WA 348). With Ivan Turgenev, he had much more affinity. Through their mutual friend, Peter Lavrov, they had met in Paris in 1878, and Kropotkin had become a great admirer of Turgenev, both as a man and as a writer: he considered Turgenev to be probably the greatest novelist of the nineteenth century, while he thought Turgenev's sketches of Russian village life in *A Sportsman's Notebook* (the title was given to mislead the censors), was important in seriously undermining the institution of serfdom (MR 408-413).

Kropotkin was also a great admirer of Leo Tolstoy (1828-1910). The two men had much in common: both were anarchists and opposed state power and capitalist exploitation; both dedicated their lives to the cause of social justice; both expressed a pantheistic sensibility towards the natural world; both had an intense interest in agrarian communes, and both thought that it was the common people, not kings and generals, who made history. Kropotkin had a deep admiration and respect for Tolstoy. He had read Tolstoy's great novel *War and Peace* several times, for it gave him indescribable aesthetic pleasure. Kropotkin suggested that Tolstoy's writings contained more truth and poetry than was usually evident in the novel form (RL 118). He also respected Tolstoy for the moral stance he had taken with regard to the Tsarist state, and for devoting his life and prestige to the cause of the oppressed in Russia - at some risk to himself (WA 351).

Although the two men never met, their relationship was always one of mutual respect, for Tolstoy had an equally high regard for Kropotkin, both for his moral integrity and for his political writings and activities. The two men differed in that Tolstoy, as a Christian pacifist, condemned unreservedly all forms of violence, and advocated 'non-resistance to evil', that is, one should never resort to force or violence in resisting oppression. This meant, for Tolstoy, that the transformation of society must come about through individual moral change, not by changing social institutions through force ,such as peasants forcefully taking possession of the land, which would inevitably provoke violence. But Tolstoy's castigation of state power and capitalism; his critiques of patriotism, institutional religion, militarism and war, and his compassionate defence of the peasant commune - all this was acknowledged and admired by Kropotkin and by other anarchists (Avrich 1967: 36).

Kropotkin was to become a close friend of Vladimir Chertkov, Tolstoy's leading disciple and his close confidant. Chertkov became a frequent visitor to the Kropotkins' home in

Bromley. As with Mavor, Chertkov often transmitted messages and greetings between the two men. On one occasion, Tolstoy wrote to Chertkov:

> Kropotkin's letter has pleased me very much. His arguments in favour of violence do not seem to me to be the expression of his opinions, but only of his fidelity to the banner under which he has served so honestly all his life. He cannot fail to see that the protest against violence, in order to be strong, must have a solid foundation (WA 351-2).

Kropotkin felt he had been misunderstood, and affirmed to Chertkov that he sympathised a great deal with the ideas of Tolstoy. He considered Tolstoy to be one of the most loved men in the world (RL 161). Tolstoy, in turn, spoke highly of Kropotkin's *Memoirs* and his book on agrarian economics, *Fields, Factories and Workshops*.

Although some of the most famous anarchists have been Russians - Bakunin, Tolstoy, Goldman and Kropotkin - anarchism as a social movement did not exist in Russia until the beginning of the twentieth century. The Anarchist Library group in Geneva had created the initial stirrings, and then in 1902, a group of Kropotkin's devotees in London had issued a translation of *The Conquest of Bread*, under the telling title of *Khleb i Volya* (Bread and Liberty). But as Avrich writes:

> Not until 1903, when the rising ferment in Russia indicated that a full-scale revolution might be in the offing, was a lasting anarchist movement inaugurated both inside the Tsarist empire and in the emigré colonies of Western Europe (Avrich 1967: 38).

For it was in that year that a small group of anarchists in Geneva who had fallen under the influence of Kropotkin, as well as of Cherkezov's personal proselytising, founded a monthly journal, also called *Khleb i Volya*. The editor, and the most active member of the group, was the Georgian anarchist Georgy Goghelia. Kropotkin gave *Khleb i Volya* his enthusiastic support. He contributed articles to the paper, mostly on Russian issues and on the impending Russian revolution. The paper was smuggled into Russia, and prior to the revolution it was the leading anarchist newspaper. It was eagerly passed around various groups of students and workers in Russia (op cit, 39). But although Kropotkin wholeheartedly supported the 'Bread and Liberty' group, he did not support its terrorist acts, although he recognised as well as anyone the brutal and repressive nature of the Tsarist regime, which suppressed all forms of political dissent.

> Despite his own dislike for violence, he did not condemn specific acts, but he would not support terrorism conceived and followed as a definite policy, since he thought that this drove any movement that practised it into conspiratorial action and so divorced it from the people (WA 357)

Violence and terrorism, for Kropotkin, was acceptable only if it was spontaneous, an act of desperation on the part of oppressed people, not as a 'calculated strategy' advocated by revolutionary leaders. Kropotkin continually warned against any flirtation with terrorism, which, as long ago as 1872, when he was a member of the Chaikovsky Circle, he had repudiated as a strategy for meaningful social change (MK 207).

From the time of the Anglo-Boer war, which he thought the most unjust war ever fought, Kropotkin's health began to decline. He tended to make fewer public appearances, and began to spend the summer months at coastal resorts, spending time at Hove, or Eastbourne, or in the Channel Islands. He had a number of serious heart attacks which almost proved fatal and he sometimes spent several weeks bed-ridden.

In the autumn of 1907 Kropotkin moved from Bromley to Highgate. There he wrote his monumental and exhaustive *The Great French Revolution* (1909). He had been gathering material for many years - his interest having been sparked in his boyhood days by his French tutor. Ever since his arrival in England in 1886 he had been working on the study. His aim in studying the French Revolution was to seek lessons, or to draw a paradigm for future revolutions. He was interested in exploring the economic and social causes of the Revolution, stressing the importance of the struggles of the common people - the peasants and workers (*sans-culottes*) - rather than focusing, as earlier historians had done, on the dramatic roles played by such figures as Robespierre and Danton. As with his other studies, Kropotkin's book on the history of the French Revolution is readable, absorbing and scholarly without ever being scholastic. It is thought to be one of the best historical studies written by an anarchist (MK 216).

But although Kropotkin had begun to feel despondent at the decline of revolutionary socialism in England during the 1890s with the general trend towards parliamentarianism, his hopes were renewed at the turn of the century by two important events. The first was the rise of revolutionary syndicalism, particularly in France, where his friend Guillaume was now active as an enthusiastic syndicalist. It was also evident in Britain, where Tom Mann attempted to sustain a revolutionary perspective within the Trade Union movement.

The syndicalist movement had emerged around 1890 and Kropotkin was always to support and to express solidarity with syndicalism, viewing it as essentially a rebirth and re-affirmation of the First International with its emphasis on working-class unity and industrial action (rather than terrorism) and in having a federalist political programme. He always wholeheartedly supported the syndicalist (trade union) movement, seeing it as of crucial importance in the struggle of working-class people against capitalism (MK 176).

The second event was the growing revolutionary situation in Russia, for in the first decade of the century the struggle against autocracy was increasing in strength and gaining popular support. In July 1902 Kropotkin remarked to Guillaume that the "history of 1789" seemed to be repeating itself in Russia, and in the following year there was a whole succession of student demonstrations, peasant revolts and strikes among the industrial workers of Moscow and St Petersburg (WA 353-4). And, as we have already noted, at this time the first anarchist groups in Russia began to emerge.

Ever since 1897, when student demonstrations had been put down with brutality and torture, Kropotkin had been writing letters and articles condemning the Tsarist autocracy, exposing its persecutions and injustices, and defending its victims. The culmination of the struggles of the Russian people came in 1905 with the first Russian revolution. One Sunday in January 1905 nearly two hundred thousand working people of St Petersburg marched to the Winter Palace. They carried a petition, intending to lay their grievances before their beloved Tsar. They sang hymns and carried crosses and icons of the Tsar.

At that time the Tsar evinced a certain mystique. He was viewed by most peasants and working people as a benevolent father, somehow not responsible for the tyranny, exploitation and poverty that the common people daily experienced. They were hoping that the Tsar might do something to improve their working conditions. Even Bakunin, when he was first imprisoned, expressed a similar and rather servile attitude towards the Tsar, when he wrote his famous *Confession* (1851).

But the demonstrators in St Petersburg that day did not receive the paternal blessings of the Tsar - they received instead a volley of bullets from the lines of infantry that had been arranged to meet them. Around one hundred and fifty people were killed and several hundred wounded: women, men and children. The events of that 'Bloody Sunday' completely severed the ancient bond between the Tsar and 'his' people; it destroyed all faith the people had in the Tsar's goodwill and benevolence.

Kropotkin considered this event not simply as an act of rebellion, but as the early stages of a revolution of the people that was similar to the revolution in France between 1789 and 1793. Throughout 1905 and 1906, he wrote a series of articles and pamphlets on the impending Russian revolution, as well as playing an active part, along with Alexander Shapiro and Marie Goldsmith (both Russian Jewish anarchists) in a number of anarchist conferences held in London and Paris. The aim of these conferences was to co-ordinate the activities of Russian exiles with the anarchist movement in Russia, and to discuss the political strategies. The major issues discussed were terrorism, constitutionalism, expropriation, and the relationship of anarchists to the emerging soviets.

At these meetings Kropotkin often spoke from a historical perspective, comparing the contemporary Russian situation with that of France in 1789. He consistently argued against terrorism. The social revolution he envisaged and advocated was to be a popular revolution, involving expropriation: 'all the land was to be seized by and returned to the people; factories and plants were to be taken over by the workers who operated them' (MK 211). But what was clear about Kropotkin with regard to strategy is that he had an absolute aversion to the idea that 'the end justifies the means'. Direct action should be the basis of struggle, not terrorism, nor parliamentary methods. Trade Union activities were important, but these should not be allowed to become the subsidiary organisation of political parties or the state. All these ideas were expressed cogently in his lectures and addresses to anarchist meetings, and published as the pamphlet *The Russian Revolution and Anarchism* (1907). Kropotkin thought at this time of returning to Russia but was advised against this by Max Nettlau.

In October 1906, along with Marie Goldsmith, Alexander Shapiro and other anarchists, Kropotkin also helped to launch a Russian emigré periodical *Listki Khleb i Volya*. Over three thousand copies were printed, although most of these were distributed in Britain and the United States rather than in Russia. At this time there were a considerable number of emigré Jewish workers and radicals settled in east London. They were inspired by a German philosopher, teacher and labour organiser Rudolf Rocker (1873-1958), who worked among them, and who has been described as 'an anarchist missionary to the Jews'. Although Rocker and the east London anarchists had strong tendencies towards syndicalism, Kropotkin was always supportive and closely associated with this group of Jewish anarchists (WA 368, Fishman 1975).

The year 1905, however, was a sad year for Kropotkin. Louise Michel died at Marseille in January; and Elisée Reclus in Brussels in July. Both were aged seventy-five, and had long been close friends of Kropotkin. He had felt particularly close to Reclus, for Reclus was not only an anarchist communist but shared his interests in geography. Kropotkin wrote long obituary articles in both *Freedom* and the *Geographical Journal*, expressing his intense love and admiration for Reclus, whom he respected both as a scientist and as a warm and engaging personality (WA 293).

As the anarchist movement began to emerge in Russia in the aftermath of the 1905 revolution, three distinct groups of anarchists became evident. The first were the anarchist communists, who generally followed the principles and ideas of Kropotkin, although many within this faction still supported acts of terrorism. The most important of these was the group known as *Chernoe Znamya* (The Black Banner - the emblem of anarchism).

Secondly, there were the individualistic anarchists, who were particularly evident in St Petersburg, Moscow and Kiev. This group was strongly influenced by the ideas of Nietzsche, in advocating the complete repudiation of all bourgeois values. But they were equally influenced by the writings of Stirner and Tucker, and were highly critical of the anarchist communism of Kropotkin, believing that even voluntary communes might limit the freedom of the individual: they exalted the ego over and above all claims of the collective. The *Beznachalie* (without authority), a group centred on St Petersburg, were close to the individualists, although they claimed to be anarchist communists (Avrich 1967: 44-56).

The third group were the anarcho-syndicalists, which included Marie Goldsmith and Georgy Gohelia, who were originally members of the *Khleb i Volya* (Bread and Liberty) group associated with Kropotkin. The syndicalists, influenced by Marx and his doctrine of class struggle, put a fundamental emphasis on the industrial proletariat. As Avrich writes, like the Russian Marxists (social democrats), the syndicalists

> placed class struggles at the centre of all things and yet - once again like the early Marxists - eschewed terrorism in favour of marshalling the workers for the approaching conflict with the bosses and government (Avrich 1967: 88).

But soon revolutionary activities in Russia began to subside as reaction set in. The Russian parliament, the Duma, was dissolved by the Tsar. Within a year the autocracy and the counter-revolution had consolidated itself. Nicholas II and his minister Stolypin embarked on a campaign of repression which included "execution, imprisonment, exile, torture, and beatings on a scale unprecedented in recent Russian history" (WA 372). In the aftermath of the 1905 revolution, acts of insurrection and terrorism became almost commonplace, usually involving radical socialist revolutionaries or individualist anarchists. In August 1906, Stolypin's own summer-house was blown up, wounding his son and daughter, and killing thirty-two people. Thus during the period of 'pacification' following the 1905 revolution, scores of anarchists were sentenced to long terms in prison or to forced labour camps (Avrich 1967: 64-70).

The brutalities and repression of Nicholas II's autocratic rule, which put an end to all civil liberties, aroused the indignation not only of socialists, but also of many liberal MPs in Britain. Kropotkin co-operated with the Parliamentary Committee on Russian Affairs in producing a detailed exposé of the infamous Russian state. The publication was entitled 'The Terror in Russia'(1909). It was, as Miller remarks, perhaps the closest Kropotkin came to collaborating with a government agency (MK 201).

In 1911 Kropotkin and Sophie moved to Brighton, finding a house at Kemp Town. Kropotkin was now sixty-nine years of age, and in declining health, for not only did he experience heart problems, but he also had bouts of bronchitis and pneumonia. He went to Brighton for health reasons, and after 1908 increasingly spent the winter months in Switzerland and Italy in order to avoid the damp English climate. Brighton was to be his last home in England. He spent little time in London, going there only for the occasional lecture or meeting. In 1912 there was a large meeting in London to celebrate Kropotkin's seventieth birthday, attended by G. B. Shaw, Lansbury and Hyndman, as well as by his many anarchist friends, but Kropotkin was too ill to attend.

After finishing his important historical study of the French Revolution, published in 1909, Kropotkin began writing a series of articles on evolutionary theory published in *Nineteenth Century* (1910-1915). These were not collected in book form until the present decade (EE). But although, through force of circumstances, Kropotkin had become something of a scholarly recluse, he still remained a devoted militant anarchist, continuing to write anarchist propaganda. Many of his political tracts and articles were published in *Freedom*.

8: The Crisis

All socialist parties, whether Marxist or anarchist, found themselves in a state of acute crisis at the outbreak of the First World War. The internationalist image of the socialist movement; its insistence on the close links between capitalist imperialism and the state, and the notion that the working class have no country, was suddenly shattered when hundreds of socialists joined the colours of their respective nation states. Kropotkin, much to the chagrin and despair of all his anarchist comrades, unequivocally came out in support of the Entente, and blamed Germany for the war. His support for the war both contradicted his own anarchist philosophy, which he had clearly spelled out in numerous tracts and articles, and, at the same time, reflected a certain continuity in his own thinking, for he had always been vehemently opposed to the German state.

Although Kropotkin was never a pacifist, he was vehemently opposed to war, both emotionally and ideologically. His classic essay on war, *La Guerre* (1882) argues that the wars at the end of the nineteenth century were essentially concerned with economic domination and were fought for the benefit of the barons of high finance and industry. Wars, he suggests, were instigated to impose customs and tariffs on neighbours; to open up new markets, and to exploit people at the periphery of capitalism. War is seen as the logical extension of the violence and greed of capitalism; for Kropotkin, the state, war and capitalism are intrinsically linked (WR 64-66).

Kropotkin had been vigorously against the Franco-Russian alliance, fearing this would lead to the restoration of the French monarchy; he had opposed both the Anglo-Boer war and the war between Japan and Russia. Appalled at the 'jingoism' that the Anglo-Boer war evoked, he had urged British workers to refuse to obey the government and to oppose colonialism. He equally refused to take sides in the Russo-Japanese war, which was not, he felt, a defensive war, but an imperialist war which was destructive to the people in both countries (MK 221-2). But Kropotkin saw the First World War quite differently, and strongly advocated supporting the Allies against Germany. His motives for this have been succinctly expressed by Avrich:

> His action was prompted mainly by the fear that German militarism and authoritarianism might prove fatal to social progress in France, the revered land of the great revolution and the Paris Commune. Germany, with its political and economic centralisation and its Junker spirit of regimentation, epitomised everything Kropotkin detested. As the bulwark of Statism, it blocked Europe's path to the libertarian society of his dreams. He was unshakeably convinced that the Kaiser had launched the war with the aim of dominating the continent (1988: 69).

Kropotkin had long held such views. More than a decade earlier, in 1899, he had written a series of articles on 'Caesarism', provoked by the Dreyfus Affair in Paris. In these articles he expressed his extreme antipathy towards the German state. He considered the defeat of France in 1870-71 as "the triumph of militarism in Europe, of military and political despotism, and at the same time the worship of the state, of authority and state socialism, which is in reality nothing but state capitalism" (WA 289). He seemed to equate both German culture and philosophy, and Marxism with the German state. Like many Russian revolutionaries, Herzen and Bakunin included, Kropotkin detested all things German, for there had been a long and close association between the Romanov dynasty and Prussian militarism. Though Herzen had a Prussian mother and Bakunin was an eager disciple of Hegel in his early years, both were, in sentiment, anti-German.

Kropotkin disliked Hegelian metaphysics and had a fervent and passionate interest in the French Revolution and in the French socialist tradition, as well as a love of France that was akin to an 'adoptive patriotism' (WA 374). There was in fact a logic and a constancy in his anti-German sentiments and his lifelong admiration for France. As Purchase writes: Kropotkin sincerely believed that the German people posed a real threat to the progress of socialism in Europe, and thus allowed himself to indulge in the 'most reckless jingoism' (1996: 33).

All this was most distressing for his anarchist friends and comrades. Although there is some truth in the notion that German imperialism and militarism were to some extent responsible for the war and the international debacle, Kropotkin's tendency to equate peoples with the state, and to think in nationalist terms, greatly upset his comrades. Many, like Errico Malatesta and Emma Goldman, were close friends and devotees. They could hardly believe that Kropotkin would abandon his internationalist outlook and his anarchist principles, especially as Kropotkin had long been advocating an anti-militarist position. But the anarchist historian Max Nettlau recorded that Kropotkin's attitude in 1914 did not surprise him, since Kropotkin 'could not have acted otherwise'. Both Nettlau and Gustave Landauer suggested that although Kropotkin in his scientific writings consulted and acknowledged German sources with interest and accuracy, he was in fact largely ignorant about German politics and cultural life (1921 (1992): 388).

Having then, at the outbreak of the war, declared himself an enthusiastic supporter of the Entente, Kropotkin wrote articles and letters urging his comrades and friends to take a stand against German militarism. In September 1914 he wrote a letter to Jean Grave in which he asked 'What world of illusions do you inhabit to talk of peace', and in the most belligerent fashion wrote of the Germans as 'savage hordes', 'an army of Huns', who were about to trample humanity underfoot. Kropotkin urged the production of cannons; expressed his admiration for the worst Allied statesmen and generals, and treated as cowards those anarchists who refused to support the war effort. He regretted that his age and poor health prevented him from fighting the Germans. Kropotkin forgot completely that he was an anarchist and an internationalist, and he forgot too, that only a short time before (1913) he had written about the impending capitalist war.

His close friend Malatesta thought Kropotkin a 'truly pathological case' and described his estrangement from Kropotkin on this issue as 'one of the saddest, most painful moments of my life' (1931 (1992): 398). Kropotkin could only think of defending his beloved France, and the revolutionary tradition he identified with it, against German aggression, fearing that the German army would impose on Europe a century of militarism.

Kropotkin was not alone in supporting the war, for Jean Grave, Varlaan Cherkezov and Charles Malato, as well as the veteran social democrat Georgi Plekhanov, all sided with the Allies. But the majority of anarchists opposed the war, and affirmed the principles of anarchism. Marxists lost no opportunity to deride Kropotkin for his chauvinism, with Trotsky noting that Kropotkin 'who had a weakness ever since youth for the Narodniks, made use of the war to disavow everything he had been teaching for almost half a century' (1980/1: 230).

In February 1915 a group of thirty-five socialists and anarchists including Emma Goldman, Alexander Berkman, Errico Malatesta, Alexander Shapiro and the Dutch socialist Domela Nieuwenhuis, issued a signed manifesto protesting against any participation in the war effort. The root of the war, this affirmed, was located exclusively in the existence of the state, and 'the state is merely oppression organised for the benefit of a privileged minority' - the capitalists. This is the case whatever the form of the state: republican, absolutist or a constitutional monarchy. The anarchist position in the present crisis was therefore to support neither side in the conflict. The manifesto declared that 'no matter where they may find themselves, the anarchists' role in the current tragedy is to carry on proclaiming that there is but one war of liberation: the one waged in every country by the oppressed against the oppressor, by the exploited against the exploiter'. True social justice could therefore only be achieved through anarchism, and war and militarism eradicated only through the 'utter demolition' of the state and its agencies of coercion (Guerin 1998: 2/35-36)j.

Throughout the early part of the war Malatesta, in the pages of *Freedom*, critiqued Kropotkin's support of the Allies, arguing that Kropotkin, whom he clearly loved and respected, had forgotten and betrayed his own anarchist principles.Malatesta did not see himself as a pacifist. 'I fight', he wrote, 'as we all do, for the triumph of peace and of fraternity, amongst all human beings'. But he bewailed the fact that many anarchists had begun to behave like bourgeois nationalists, and to think of 'France' and 'Germany' as if they were 'homogenous ethnographic units', each having its own interests and mission. As these were now locked in a historic struggle, there was the issue of which side to support. But to think along these lines, Malatesta argued, was to forget socialism, the class struggle and the internationalist perspective of anarchism - for anarchists had always fought against patriotism. It is the duty of anarchists, he stressed, to support neither side in the conflict, but rather

> to do everything that can weaken the state and the capitalist class, and to take as the only guide to their conduct the interests of socialism.

Malatesta goes on to suggest that in his opinion:

the victory of Germany would certainly mean the triumph of militarism and of reaction, but the triumph of the Allies would mean a Russo-English (ie a knouto-capitalist) domination in Europe and Asia, conscription and the development of a militarist spirit in England, and perhaps monarchist reaction in France.

He had, he admitted, no more confidence in the bloody Tsar, or in the British state which crushed the Boer Republics, or the French bourgeoisie, who massacred the people of Morocco, than he had of the German state (1992 (1914): 389-92).

Malatesta found it painful and difficult to challenge Kropotkin. But when in February 1916 Kropotkin and his supporters (Grave, Malato, Cherkezov) issued a manifesto in support of the war, Malatesta was again induced to protest against the 'pro-government anarchists'. Although not doubting Kropotkin's good faith and good intentions, Malatesta stressed the need for all anarchists to remain true to their principles, and to disassociate themselves from the 'self-styled anarchists' who co-operate with and support governments and the capitalist classes of specific countries (1992 (1916): 393). He continued to urge Kropotkin to repudiate his pro-war position and to re-affirm his own anarchist principles. But Malatesta admitted that he under-estimated the force of Kropotkin's anti-German prejudices and the patriotic fervour of his attachment to France (WA 382).

Although Thomas Keell, the editor of *Freedom,* tried to mediate, Kropotkin's pro-war stance created a split within the anarchist movement, even though the great majority of anarchists were opposed to the war. Besides Malatesta, these included Berkman, Shapiro, Rocker, Goldman, Lilian Wolfe, Saul Yanovsky, Luigi Fabbri and many others. Indeed, it was evident that the great majority of anarchists were opposed to Kropotkin, and Kropotkin's connection with *Freedom* came, after an acrimonious exchange, to an abrupt end in January 1915. After a close association with the paper for almost three decades and having been a regular contributor to its pages during all these years, Kropotkin ceased to be associated with *Freedom* and never again wrote for the journal.

During much of 1915 and 1916, while living in Brighton, Kropotkin, now in his seventies, was not in good health. He had two operations on his chest, and spent much of his time in a wheelchair. But although he kept up with his extensive correspondence, Kropotkin became increasingly isolated from the main anarchist movement, and particularly felt his estrangement from Malatesta. He lost contact with many of his anarchist comrades, many of whom were devotees inspired by his earlier writings. He still kept in touch with Shapiro, Goldman and Rocker, and a few trade unionists and scientific or literacy friends who came to visit him in Brighton, but to all extents and purposes 'it looked like a dull end to an active life' (WA 387). Then, quite unexpectedly in February 1917, there was revolutionary insurrection in Russia, and Kropotkin's life was suddenly transformed.

9: The End of Exile

Although Kropotkin had long anticipated and dreaded the outbreak of the First World War, he was quite taken by surprise by the Russian Revolution of February 1917. Yet it was, in many ways, the kind of spontaneous revolution of the working people that he had long envisaged and advocated. It seemed that the moment had come for which he and many other Russian revolutionaries had long been struggling, namely the downfall of the Tsarist regime. The Revolution also enabled Kropotkin to realise the dream of his life: to return to his native Russia and to end his forty years of exile. At the end of 1916 he had written to his friend, the Russian historian Sergei Melgunov, that a return to Russia would be the 'greatest day of my life' (MK 232), and now the opportunity had come.

Soon after hearing the news of the February Revolution and of the Tsar's abdication, Kropotkin and Sophie therefore decided to return to Russia, and began making preparations. Kropotkin felt obliged to write a farewell letter to the working people of Western Europe, which he gave to the trade-union leader John Turner. Kropotkin wrote:

> After having worked in your midst for forty years, I cannot leave Western Europe without sending you a few words of farewell. From the depth of my heart I thank you for the reception - more than fraternal - that I have found in your midst. The International Workingmen's Association was not for me a mere abstract world. Amidst the working men of Switzerland, France, Britain, Spain, Italy, the United States, I was in a society of brothers and friends. And in your struggles, each time I had the opportunity to take part in them I lived the best moments of my life.

Kropotkin insisted that the workers of Europe should continue to support the war against German aggression and militarism, and urged the worker-control of industry. The workers, he wrote, 'the producers, must become the managers of producing concerns' (WA 394).

The letter to the workers of Western Europe was published in *Freedom* in July 1917, and it carried with it the editorial comment:

> In bidding farewell to Kropotkin we can but hope that by contact with the Russian workers he may realise the errors of his attitude on the war, and with them work in the building of an anarchist society of which he was such an enthusiastic exponent prior to the war. His numerous anarchist books and pamphlets will be read and remembered long after his patriotic backsliding in this war has been forgotten (MK309)

Travelling incognito, under the name of Sergei Tiurin (one of his Russian friends), Kropotkin sailed from Aberdeen in June 1917. But in both Norway and Sweden he was greeted by welcoming demonstrations, and en route to Russia, he attended several receptions, as well as meeting with the Swedish social democrat Hjalmar Branting to discuss the war situation. In a memorandum to Branting, Kropotkin outlined his views for a peace settlement at the end of the war: it advocated the return of Alsace and Lorraine to France, and the recognition of Poland and Serbia as independent states. As Miller writes:

> The irony of an anarchist proposing the establishment of new states was another aspect of his war position

that continued to stun his followers and anger his critics (MK 234).

Kropotkin arrived by train in Petrograd on June 12, 1917, and though it was only two o'clock in the morning, a crowd of around sixty thousand people came to welcome him back to Russia. Through his writings, Kropotkin had become something of a legendary figure. Because Kropotkin had taken a 'patriotic' stand on the war, two ministers from the provisional government - Kerensky and Skobolev - had also come to meet him. But it is significant that there were no anti-war anarchists there to greet him at the station.

Kropotkin and Sophie took up residence in Moscow. It was evident from the outset that, politically, Kropotkin cut a rather lonely figure. Most of the Russian anarchists kept aloof from him, feeling that, by supporting the war, Kropotkin had betrayed his own anarchist principles. He was out of favour, too, with the social revolutionaries and the social democrats, with the Marxists generally dismissing him as an 'old fool' or as 'petty-bourgeois'. A meeting with an old friend, Ivan Knizhnik (who had become a Marxist), soon after his arrival, ended in bitterness and acrimony, as Kropotkin remained vehemently anti-Marxist.

But Kropotkin's relationship with those who supported the war, most of whom were liberals in favour of some sort of constitutional government, was equally problematic. He was on friendly terms with the Prime Minister Kerensky, and had several meetings with Prince Lvov, for Kropotkin sympathised with Lvov's efforts to strengthen the autonomy of the provincial (*Zemstvo*) administration. But Kropotkin adamantly refused to accept any government position, as well as declining a state pension or a lodging in the Winter Palace. In August 1917, however, he participated in the Moscow National Conference called by Kerensky, the essential purpose of which was to rally the various political parties and groups behind the provincial government. Along with the Marxist Plekhanov, who also took a pro-war stand, Kropotkin was one of the key speakers of the conference. In his speech Kropotkin strongly advocated continuing the war effort against Germany, and seemed to equate Russia and the revolution, implying that a defence of the country was equivalent to a defence of the revolution. But he also proposed to the meeting that Russia should declare itself a republic – 'we need a federation such as they have in the United States' - and that the industrial bourgeoisie should do all in its power to alleviate the suffering and hardships of working people, and share their 'knowledge' of commerce and industry.

Although Kropotkin's speech received a long and warm ovation, it smacked of compromise with the provisional government, and of course allowed the Marxists to ridicule and discredit not only Kropotkin but the whole anarchist movement. Trotsky noted how bizarre it was that the apostle of non-government should give his support to the right-wing elements of the conference- the landlords, industrialists and generals - all in the defence of the Motherland (1980: 178-79).

This speech also caused an ever-deepening rift between Kropotkin and the anarchist movement in Russia, and indicated the degree to which Kropotkin - presented to the conference as the 'icon of the Russian revolution' - was out of touch with the rising forces of the Revolution. For the soldiers at the front, the working classes and the peasants - all longing for bread and peace - were beginning to take the situation into their own hands. It has been stressed that Kropotkin 'failed to perceive the real popular forces behind the October movement' (WA 397-402, MK 236). It seems that Kropotkin's intense hatred of Prussian militarism completely distorted his judgement: as Emma Goldman put it, he seemed to justify all measures to crush the 'Prussian menace' (1931/2: 564). No wonder his support of the Entente, his advocacy of a federal republic, and his 'patriotic' compromises with the Russian bougeoisie were all a source of embarrassment to most anarchists. When Emma Goldman learned that Kropotkin, 'the anarchist, humanitarian and gentlest of beings' - had taken sides with the Allies, she found it hard to believe, thinking at first that it was a hoax. She responded immediately by publishing Kropotkin's article 'Capitalism and War' in *Mother Earth*, for it embodied a logical and a convincing refutation of Kropotkin's own present position (1931/2: 564-5).

Little is known of Kropotkin's activities during the October Revolution in 1917, although he is alleged to have said to Alexander Atabekian, the Armenian anarchist, that 'this buries the revolution'. For although the events of October had followed the anarchist pattern of a popular revolution - workers had assumed control of the factories, peasants had taken over the land, and soldiers had been expressing their hatred of the war by widespread desertion - the outcome of this revolution was the seizure of power by the Bolsheviks. What emerged from the social revolution was a Marxist revolutionary government. As Miller writes:

> Next to being overcome by Russian militarism as France had been in 1871, state socialism was perhaps the most horrible misfortune Kropotkin could have imagined for Russia (MK 237).

While still living in Moscow in the early months of 1918, Kropotkin became involved with the activities of the Federalist League. This was a group of scholars who were interested in sociological issues, and who wished to promote federalism and decentralised politics. They usually met at Kropotkin's home and planned a series of studies of federalism in all its aspects, for Kropotkin saw this work as a means of countering both the state centralism of the Bolsheviks and any possible Monarchist restoration. Although the League was primarily involved in scientific work, it was, however, soon suppressed by the Bolsheviks.

In June, Kropotkin was visited by the Ukrainian anarchist Nestor Makhno (1889-1934), who sought Kropotkin's advice on what kind of political strategies would be appropriate in the Ukraine. This visit was a kind of pilgrimage for Makhno, and in his memoirs he speaks highly of Kropotkin and the importance of their meeting. Makhno was later to become famous as a kind of anarchist Robin Hood, as Guerin describes him (1998/2: 123), when he became the leader of the guerrilla bands of Ukrainian peasants who not only fought against the occupying Austro-German armies and the counter-revolutionary forces, but also against the Bolshevik Red Army commanded by Trotsky (Guerin 1998/2: 123-62, Arshinov 1974).

Kropotkin was unhappy living in Moscow. Physically weak as he was, and hardly able to walk, city life was not conducive to his health, while the apartment in a block of flats was cold and depressing. Kropotkin was also greatly perturbed at the increasing authoritarianism of the Bolshevik regime: to an American visitor he described the Bolsheviks as a group of 'robbers and gangsters, set upon looting and destruction' (WA 406). He and Sophie therefore decided, in June 1918, to move to the small village of Dmitrov, forty miles north of Moscow.

Here Kropotkin virtually became an 'internal emigré', cut off from the political life of the capital; his correspondence curtailed by a disorganised postal service. He was increasingly prevented by ill health from writing. In the last three years of his life he wrote only a few letters, a couple of short, unfinished essays, and a work on *Ethics*. This work, however, he never completed, and the unfinished text would be published posthumously in 1924.

But significantly, in Kropotkin's last years, he began to draw closer to his anarchist friends, in spite of his support for the war. Unlike him, they had supported the October Revolution. But now they found themselves among the first victims of Bolshevik repression (WA 407).

In April 1918 an armed force of some five thousand troops, acting under the orders of the government, smashed all the anarchist organisations in Moscow. This was done under the pretext of eradicating the 'banditry' in the anarchist ranks. After the attempt on Lenin's life in August 1918 by the socialist Dora Kaplan, an even greater 'bacchanalia of terror' was unleashed upon the Russian people - with the shooting and imprisonment of all potential dissidents to the 'dictatorship' of the Bolshevik party under Lenin (Maximoff 1979: 57-59).

There has been a tendency to minimise the role that anarchists played in the Russian Revolution, but almost everywhere in Russia in 1917 and 1918 there arose anarchist groups, movements and tendencies. Some were of slight import and ephemeral, but in spite of their ideological differences (ranging from individualism and syndicalism to anarchist communism), most anarchists had fought alongside the Bolsheviks in the October Revolution. The two most active groups were the Union of Anarcho-Syndicalist Propaganda, which bore the name *Golos Truda* (The Voice of Labour), and the Federation of Anarchist Groups in Moscow, who published the daily newspaper, *Anarkhiya* (Anarchy) and who were of anarchist communist persuasion. *Golos Truda* was said to have rivalled Lenin's *Pravda* in its influence. But all these groups eventually met the same fate: 'brutal suppression by the "Soviet" authority' (Voline 1974: 267-69).

At the end of the war, Kropotkin found himself in a rather ambivalent position with regard to the October Revolution. On the one hand, he strongly supported the Revolution, but feared a reaction and the restoration of the monarchy. On the other hand, he was highly critical of the authoritarian nature of the Bolshevik revolutionary government.

His position is made clear in letters he wrote around 1919. In one that he wrote to the Danish literary critic Georg Brandes (1842-1927), Kropotkin drew similarities between the Russian and French Revolutions and suggested that "we are now at the same stage as France was during the Jacobin revolution from September 1792 to July 1794, except that now a social revolution is seeking expression". Although the French Jacobins put an end to feudalism and advocated the equal rights of all citizens, their dictatorial methods were to be deplored. A similar situation, Kropotkin argued, existed in Russia, for the Bolshevik Party dictatorship is

> striving to introduce the socialisation of land, industry and commerce. This change which they are striving to effect is the fundamental principle of socialism. Unfortunately, the methods by which they seek to establish communism like Babeuf's in a strongly centralised state makes success absolutely impossible and paralyses the constructive work of the people (SW 320).

This situation inevitably nourished a potentially dangerous reaction, and with the Western powers advocating armed intervention to re-establish 'order' in Russia, Kropotkin was clearly troubled. He wrote to Brandes: 'I protest with all my strength against any type of armed intervention by the Allies in Russian affairs. This intervention would produce an outburst of Russian chauvinism'. It would lead, he felt, only to reaction, the restoration of the monarchy and rivers of blood (SW 321).

In April 1919 he wrote a letter expressing similar views to the workers of Western Europe, emphasising the inherent evils of a one-party dictatorship. He wrote:

> I have to tell you candidly that, in my view, this attempt to erect a communist republic upon a base of strongly centralised state communism, under the iron law of a one party dictatorship is heading for fiasco. We in Russia are beginning to learn how communism should not be introduced (Guerin 1998/1: 282).

Kropotkin thoroughly approved of the idea of Soviets, that is, of workers' and peasants' councils, advocated in the revolutions of both 1905 and 1917, , and thought it a 'grand idea'. But unfortunately, under the one-party dictatorship of the Bolsheviks, the Soviets had lost all social significance, and along with the trade unions and local co-operatives, had either been suppressed or turned into adjuncts of the state bureaucracy. What had thus happened, as both Bakunin and Kropotkin had suggested in their writings on revolutionary government, was that the Bolshevik party - as the dictatorship of the proletariat - had created a powerful state bureaucracy, and an apparatus of repression that had stifled all the popular forces of the revolution.

Kropotkin concluded his letter to the workers of Western Europe by emphasising the need for a new social order based on local freedom and initiatives, and the need also to revive the notion of an International of all the world's workers, based on federal principles, not on the 'unity' of a party, as with the Second International (Guerin 1998/1: 284, MK 241).

At this time, although in his seventy-eighth year, Kropotkin seems to have remained remarkably optimistic, suggesting that the Bolsheviks would eventually come to grief through their own mistakes and misguided practices. In May 1919 Kropotkin had a meeting with Lenin. It was hardly a meeting of equals. Lenin was then at the height of his powers, supreme ruler of the Soviet state; Kropotkin was old, physically weak, and hardly in a position to influence events in Russia.

The meeting was arranged by a close associate of Lenin, Vladimir Bonch-Bruevich, who knew Kropotkin. What motivated Lenin to seek a meeting with Kropotkin is unclear. He evidently admired Kropotkin's book on the French Revolution, and may well have wanted to secure Kropotkin's goodwill and support in the difficult times through which the Bolsheviks were then passing (WA 416). Their discussion ranged over many subjects, but Kropotkin seemed eager to stress the importance of voluntary co-operatives - both in Russia and elsewhere in Western Europe, and their revolutionary significance. He was thus critical of the Bolshevik stress on bureaucratic authority, and the fact that the party had appeared to have become 'intoxicated with power'. Talking about co-operatives, however, seemed to Lenin to be nothing but 'idle chatter'; what he felt was needed was a revolution brought about by 'red terror', one which would bring down the capitalist world.

The two men clearly had very different conceptions of the revolution: for Lenin it entailed state power and the authority of the Bolshevik party. The meeting ended with Lenin emphasising the need to re-issue Kropotkin's book on the French Revolution. Kropotkin was quite happy with this suggestion, as long as it was not published by the state (SW 325-332).

In the following year Kropotkin wrote two letters to Lenin. In the first he pleaded on behalf of the employees of the government postal service, who received such meagre wages that they lived in desperate circumstances, and were literally starving. In the second, he strongly protested at the Bolshevik government's decision to hold social revolutionaries and dissident nationalists as 'hostages'. This barbaric act, he argued, represents a return to medievalism. Although Kropotkin acknowledged that the October Revolution had brought about many progressive changes in Russia, the progress of the Revolution was being severely curtailed by the Bolshevik party. One thing is indisputable, he wrote to Lenin: 'even if the dictatorship of the party were an appropriate means to bring about a blow to the capitalist system (which I strongly doubt) it is nevertheless harmful for the creation of a new socialist system. What are necessary and needed are local institutions, local forces; but there are none, anywhere'. For the Bolsheviks had suppressed all local initiatives and it is 'party committees not the Soviets, who rule Russia' (SW 336-37).

Living in Dmitrov, Kropotkin's position became similar to that of Tolstoy during his last years under the Tsarist regime: a lonely figure who seemingly represented the spirit of the Revolution. Many of the old friends from whom Kropotkin had become estranged during the war years began to re-establish contact with him. During 1920 a large number of his anarchist friends came to visit him in Dmitrov, their visits almost taking the form of a pilgrimage. They included Alexander Shapiro, Emma Goldman, Alexander Berkman, Alexander Atabekian and Nicholas Lebedev. Particularly important were the visits of two Russian anarchists especially active at that period and who came to write important studies of the Russian Revolution and its subsequent terror: Gregory Maximov (1893-1950) and Voline (Vsevolod Eichenbaum) (1882-1945).

Although the Kropotkins, living in Dmitrov, were better off than many Russians, their life was generally one of hardship, and rather spartan, especially during the winter months. It was difficult to get many provisions, and essential things, like Kerosene oil for lamps, were generally in short supply. Firewood, too, was difficult to obtain, owing to transport costs. But the Kropotkins continued, as they had done throughout their life together, to maintain a vegetable garden. Sophie undertook most of the work, as Kropotkin had become rather feeble. Emma

Goldman recalls that although Kropotkin often dug the soil, it was Sophie - herself a professional botanist - who was the real expert and instrumental in providing the household with cabbages, potatoes and other vegetables. They also kept a cow. Goldman notes in 1920 that Kropotkin was a 'mere shadow' of the sturdy man she had known in Paris and London in 1907 - although she found him in moderate health and in buoyant spirits. And he talked vividly to her of the need to combine anarcho-syndicalism with co-operatives in order to save the Revolution from the fatal blunders and fearful suffering that Russia was then experiencing under the Bolsheviks (1970/2: 769, 863-4).

In his last years Kropotkin devoted himself to two projects. One was the writing of his book, *Ethics*, in which he explored the need to establish a purely human ethics, which was of crucial importance in laying the foundations for a morality that was free of religion. The other was his practical work with the local Dmitrov Co-operative Society, for Kropotkin came increasingly to see the Co-operative Movement, being a voluntary and non-governmental form of economic activity, as essential in the formation of a new communist society. It was, as Miller perceptively writes 'a characteristic and highly appropriate end to his half-century of populist and anarchist activities that he at last lived and found meaning in the Russian countryside, in a rural town surrounded by peasants and workers' (MK 242).

In December he was visited by the Spanish trade-unionist Vilkens, who recorded that, in spite of his years, Kropotkin's thinking had retained all its lucidity, and he continued to vent his misgivings regarding the Bolshevik regime. He continued to advocate support for the Revolution, while at the same time denouncing the Bolsheviks, both for their Jesuitical spirit and their autocratic and centralist politics. As Goldman had noted, Kropotkin always clearly distinguished between the Revolution and the regime (Guerin 1998/1: 285-89).

Only two months later, on February 8th, 1921, Kropotkin died, stricken with pneumonia. Goldman, along with Alexander Shapiro, arrived at Dmitrov in the early morning, having travelled by train from Petrograd in a blizzard. She did not arrive until around four in the morning, shortly after Kropotkin had breathed his last, but was there to comfort the grief-stricken Sophie.

An enormous funeral was arranged by Kropotkin's anarchist comrades, prominent among them being Goldman, Shapiro, Berkman and Alexander Atabekian. They refused the offer of a state funeral, but, in complicated circumstances, managed to obtain the temporary release of the many anarchists who had been incarcerated in Moscow prisons by the Bolsheviks, in order that they could attend the funeral.

On the day of the funeral a procession of around a hundred thousand people followed the coffin on its five-mile journey to the cemetery of the Novo-Devichii Monastery, where Kropotkin was buried. In the procession were people from many different political persuasions, but prominent were the Anarchist groups carrying their black flags, and the banners acclaiming 'where there is authority there is no freedom'. Many tributes were made to Kropotkin at the graveside, the last being by Aaron Baron who had temporarily been released from prison. As Victor Serge records, Baron, bearded and emaciated, cried out in protest against the new despotism and the trampling of the Revolution underfoot by the Bolsheviks. This was the last great demonstration against the Bolshevik tyranny (Serge 1963: 124, WA 436).

A Kropotkin museum was established in the old house in Moscow where Kropotkin had been born, but on Sophie's death in 1938 the museum was suppressed and its contents dispersed.

Kropotkin's legacy, however, lives on in his writings, and in the influence of his ideas which continue to guide later generations.

Chronology of Important Events in the Life of Peter Kropotkin

1842	Born December 9th in Moscow.
1846	Death of his mother aged only 35.
1848	Father remarries; Revolutions in Paris, Dresden and Vienna.
1855	Reign of Tsar Alexander II begins.
1857	August: K enters the Corps of Pages, an elite Military School in St Petersburg.
1861	Emancipation of the serfs.
1862	K graduates from the Corps of Pages and begins military service in Siberia.
1863	K's first exploratory trip along the Amur River: Polish insurrection.
1865	Further travels along Amur River and to Manchuria.
1866	June: visit to Lena Gold Mines.
1867	K resigns from military service and returns to St Petersburg. He attends the University, and works as a civil servant.
1868	K elected member of the Russian Geological Society.
1871	Father dies; Paris Commune.
1872	February-May, K goes Switzerland and makes his first contacts with the Jura Federation. On his return to St Petersburg, K joins the Chaikovsky Circle of Revolutionary Populists.
1873	K drafts a manifesto for the Chaikovsky Circle.
1874	March: K arrested by police in St Petersburg and imprisoned in Peter and Paul Fortress
1876	K is moved to a prison attatched to military hospital in St Petersburg. June: he escapes from prison, arriving in England via Scandinavia.
1877	Attends Congress of the International in Verviers, and renews contacts with the Jura Federation. Settles in La Chaux-de-Fonds, Switzerland.
1878	Marriage to Sophie Ananiev, daughter of a Polish Jew, in Geneva.
1879	Begins publication of Anarchist journal *La Révolte*.
1881	July: K attends International Anarchist Congress in London. August: he is expelled with his family from Switzerland and settles in England. Tsar Alexander II is assasinated.
1882	K moves to Thonon in France.

1883	K is arrested and brought to trial in Lyons for anarchist activities. Imprisoned in Clairvaux.
1885	While in prison, K's first book is published: *Paroles d'un Révolté*.
1886	January: K is released from prison and settles in London. July: News of his brother Alexander's suicide while in exile in Siberia. October: K helps to establish the *Freedom* newspaper.
1887	Birth of K's daughter Alexandra; publication of *In Russian and French Prisons*.
1890	K begins to publish series of articles in the *Nineteenth Century* magazine on mutual aid.
1897	Visit to Canada and the United States, meeting many anarchists; K begins publication of his memoirs, which are serialised in the *Atlantic Monthly*.
1899	Publication of *Memoirs of a Revolutionist*, and *Fields, Factories and Workshops*.
1901	Second trip to the United States. K lectures in Boston on Russian literature (published as *Russian Literature: Ideals and Realities*, in 1905).
1902	Publishes *Mutual Aid: A Factor of Evolution*.
1905	Publishes articles on the Russian Revolution.
1906	Publication of the English translation of *The Conquest of Bread* (originally published in Paris in 1892).
1909	Publishes *The Great French Revolution*.
1911	K and family move to Brighton.
1914	K supports the Allies at the outbreak of the First World War: K's pro-war stand alienates many of his anarchist friends.
1917	After February Revolution, K and family return to Russia. He settles in Moscow. March: Abdication of Tsar Nicholas II.
1918	K moves to Dmitrov, about forty miles north of Moscow.
1919	Meeting with Lenin.
1920	K organises co-operatives in Dmitrov. Letters to Lenin.
1921	K dies February 8th: he is buried in Moscow.

Bibliography

Initials in brackets at the end of some entries indicate works referred to in this way in the text.

Arshinov, P	1974 *History of the Makhovist Movement* Detroit: Black & Red
Avrich, P	1967 *The Russian Anarchists* (1978 edition) New York: Norton
	1988 *Anarchist Portraits* New Jersey: Princeton University Press
Berdyaev, N	1960 *The Origin of Russian Communism* Ann Arbor: University Michigan Press
Berlin, I	1978 *Russian Thinkers* London: Hogarth Press
Broido, V	1978 *Apostles and Terrorists* London: Temple Smith
Cahm, C	1989 *Kropotkin and the Rise of Revolutionary Anarchism 1872-1886* Cambridge: Cambridge University Press
Cole, GDH	1954 *A History of Socialist Thought. Volume Two: Marxism and Anarchism 1850-1890* London: MacMillan
Cole, M	1961 *The Story of Fabian Socialism* London: Heinemann
Figes, O	1996 *A People's Tragedy* London: Random House
Fishman, WJ	1975 *East End Jewish Radicals 1875-1914* London: Duckworth

Goldman, E	1931	*Living my Life* (1970 edition) 2 Vols New York: Dover Publishers
Guerin, D	1998	*No Gods, No Masters: An Anthology of Anarchism* 2 vols Edinburgh: A K Press
Kedward, R	1971	*The Anarchists* London: MacDonald
Kitz, F	1912	*Recollections and Reflections* (Re-issue C Slienger) London: Freedom Press
Kropotkin, P	1885	*Words of a Rebel* (1992 edition) Montreal: Black Rose Books (WR)
	1887	*In Russian and French Prisons* (1991 edition) Montreal: Black Rose Press
	1899	*Memoirs of a Revolutionist* New York: Grove Press (MR)
	1906	*Conquest of Bread* (1972 edition) London: Allen Lane
	1899	*Fields, Factories and Workshops* (1974 edition) London: Freedom Press
	1902	*Mutual Aid: A Factor of Evolution* London: Heinemann
	1905	*Russian Literature: Ideals and Realities* (1991 edition) Montreal: Black Rose Books (RL)
	1909	*The Great French Revolution* (1989edition) Montreal: Black Rose Books

Kropotkin, P	1924	*Ethics: Origin and Development* Dorchester: Prism Press
	1970	*Selected Writings on Anarchism and Revolution* (ed. Martin Miller) Cambridge, Mass.: Mit Press (SW)
	1988	*Act for Yourselves* (ed. N Walter & H Becker) London: Freedom Press (AY)
	1993	*Fugitive Writings* (ed. G Woodcock) Montreal: Black Rose Books
	1995	*Evolution and Environment* (ed. G Woodcock) Montreal: Black Rose Books
Lane, J	1887	*An Anti-Statist Communist Manifesto* (1978 edition, ed. N Walter) Sanday, Orkney: Cienfuegos Press
Lebedev, NK	1932	*Perepiska: Kropotkin's Letters to his Brother Alexander* 2 vols Moscow, Leningrad: Academie
Malatesta, E	1965	'Peter Kropotkin: Recollections and Criticisms of an old Friend' (in V Richards, ed.: *Malatesta: His Life and Ideals*, London: Freedom Press pp 257-68)
	1992	'Anarchists have forgotten their principles of pro-government' (in *The Raven*, 5/4: pp389-95)
Maximoff, GP	1979	*The Guillotine at Work* Sanday, Orkney: Cienfuegos Press
Miller, MA	1976	*Kropotkin* Chicago: University Chicago Press (MK)

Morris, B	1993	*Bakunin: The Philosophy of Freedom* Montreal: Black Rose Books
Morton, AL	1973	(ed.) *Political Writings of William Morris* London: Lawrence & Wishart
Nettlau, M	1921	*'Peter Kropotkin at Work'* (in *The Raven*, 1992: 5/4: pp379-88)
Oliver, H	1983	*The International Anarchist Movement in Late Victorian London* London: Croom Helm
Purchase, G	1996	*Evolution and Revolution* Petershaw, NSW: Jura Books
Quail, J	1978	*The Slow Burning Fuse* London: Paladin Books
Serge, V	1963	*Memoirs of a Revolutionary* (trans. P Sedgwick) Oxford: Oxford University Press
Slatter, J	1981	*Kropotkin's Papers* (*Geographical Magazine*, 53: pp 917-21)
Stepniak (Kravchinsky, S)	1883	*Underground Russia* London: Smith, Elder
Trotsky, L	1980	*The History of The Russian Revolution* New York: Pathfinder Press
Utechin, S.V.	1963	*Russian Political Thought* London: Dent
Venturi, F	1960	*Roots of Revolution* Chicago: Chicago University Press
Voline	1974	*The Unknown Revolution 1917-1921* Detroit: Black & Red

Walicki, A   1980 *A History of Russian Thought*
                    Oxford: Clarendon press

Wilson, C   1979 *Three Essays on Anarchism*
                    Sanday, Orkney: Cienfuegos Press

Woodcock, G   1950 *The Anarchist Prince* (1971 edition)
Avakumovic, I   New York: Schocken Books (WA)

Yarmolinsky, A   1957 *Road to Revolution: A Century of Russian
                    Radicalism* New Jersey: Princeton University Press

Index

ANANIEV, Sophie (Kropotkin's wife): 53-4, 55, 58, 60, 65, 84, 93, 103-5

ANARCHISM: 17-18, 29, 38, 50, 52, 53, 77-8, 82-4
  In London: 60-62
  In USA: 73

ANARCHIST COMMUNISM: 52, 62, 67, 82-3

ANARCHISTS: 55, 61-5, 67, 71, 89-92, 99, 103, 105

ANARCHIST PUBLICATIONS: 54-5, 62, 63, 64-5, 71, 74, 78, 82

ARISTOCRACY IN RUSSIA: 1-2, 17, 20, 22, 34, 39, 74

BAKUNIN, Mikhail: 13, 24-6, 27, 29-31, 35, 36, 39, 44, 46, 53, 63, 74, 80, 87

BAKUNINISTS, 27, 28, 49-50, 30, 52

BOLSHEVIKS; 34, 36, 97, 98-9, 102-5

CAFIERO, Carlo: 52, 53

CAPITALISM: 16, 37, 38, 40, 86, 89, 102

CHAIKOVSKY, Nikolai: 32, 40, 48, 57, 60

CHAIKOVSKY CIRCLE, THE: 40-44...

CHERKEZOV, Valaam: 74, 78, 89, 91

CHERNYSHEVSKY, Nikolai: 23, 36, 37, 38-40, 42, 43, 47

CHERTKOV, Vladimir: 76-7

COHEN, Joseph: 57

COLLECTIVISM: 26, 29, 52

COMMUNE, PARIS: 21, 26, 27, 30, 39, 51, 74, 87

COMMUNES, PEASANT: 26, 36, 37, 38, 45, 49, 51, 57, 76, 83
See SOCIALISM: AGRARIAN SOCIALISM

COMMUNISM: 26, 52, 67, 100, 101
    Anarchist Communism: 52, 62, 63, 64, 65, 66, 67, 83, 99
    After Russian Revolution: 101, 104

CORPS OF PAGES: 5-8, 12, 47

DARWIN, Charles: 8, 18

DUMARTHERAY, François: 54-5

FABIAN SOCIETY: 61-4, 70

FEDERALISM: 28, 30, 31, 40, 52, 54, 96, 97-8 See BAKUNIN

FINLAND: 19, 44, 46, 48

FIRST INTERNATIONAL: 24, 25, 29, 50, 79 See INTERNATIONAL WORKINGMEN'S ASSOCIATION

FIRST WORLD WAR: 86-89, 91-2, 96, 100

FORWARD! Journal: 39, 64-5, 71, 90-92, 94

FOURIER, François: 34, 37, 38

FRENCH REVOLUTION: 79, 81, 85, 88, 89, 100, 102

GEOGRAPHICAL JOURNAL: 16, 82

GEOGRAPHICAL SOCIETY, RUSSIAN: 19, 21, 24, 44, 46, 47

GOETHE, Wolfgang: 7, 8, 9

GOLDMAN, Emma: 72, 73, 77, 88, 89-90, 91, 92, 103-5

GRAVE, Jean: 52, 59, 67, 71, 74, 88, 89, 91

GUILLAUME, James: 29, 31, 48, 51, 53, 54, 79

HERZEN, Alexander: 10, 36-7, 38, 39, 43...87

HERZIG, George: 54-5

HUMBOLDT, Baron Alexander von: 8, 9, 12

HYNDMAN, Henry: 57, 60, 61, 62, 84

INDEPENDENT LABOUR PARTY: 70

INTERNATIONAL WORKINGMEN'S ASSOCIATION: 24, 26, 28, 49-50, 93 ..See FIRST INTERNATIONAL

JURA FEDERATION: 24, 28, 29, 49-54 passim, 64

KARAKOZOV, Dimitri: 20, 35, 41, 46

KHLEB I VOLYA (Conquest of Bread): 77, 78, 83

KITZ, Frank: 56, 61-2, 67

KLEMENTS, Dimitri ('Kelnitz'): 20, 40, 42, 49

KROPOTKIN, Prince Alexsei Petrovich (Kropotkin's father): 1-4, 5, 6, 12, 19, 21-2, 23

KROPOTKIN, Alexander (Kropotkin's brother): 3, 4, 5, 7-8, 16-17, 19, 31, 39, 46, 48, 60

KROPOTKIN, Ekaterina (Kropotkin's mother), née Sulima: 3, 23

KROPOTKIN, Helene (Kropotkin's sister): 3, 4, 7, 46

KROPOTKIN, Nicholas (Kropotkin's brother): 3

KROPOTKIN, Pyotr Alexeivich (Peter):
Chronology:
   Birth and childhood: 1-6
   Education & tutors: 4, 6, 6-7, 25 See CORPS OF PAGES
   Military service: 13-18, 19
   Explorations in Siberia, Manchuria; Finland: 12-18, 19
   Civil Service: 20-23
   University studies: 19
      Visits Switzerland: 24-33
      Return to St Petersburg: 33-51
      Imprisoned: 44-8
      Marriage: 53-4 See ANANIEV, Sophie
      Exile: in England: 48-49, 57, 59-94
             in Switzerland: 49-56
             in France: 56-57; arrest: 58-59
      Visits USA: 72, 74
      Returns to Russia: 95
      Death: 105
Character and interests:
   Character: 53, 57, 64, 68-9
   Health: 47, 48, 78-9, 84, 92, 98, 105
   Philosophy: 7, 8-9, 18, 21, 23
As geographer: 7, 12, 14, 16, 17-21, 49, 82
As anarchist: 17-18, 39, 51-2, 53, 58, 63, 66, 68, 70, 71
As writer: 6, 18, 44-5, 49, 55, 58, 66, 67-8, 72-3, 74-5, 79, 85, 98, 104
Love of natural world: 5, 7, 9, 18, 35, 76
Relationships with peasants: 4, 5, 14, 16, 17, 20, 22, 23-4, 42, 76, 101
Relationships with workers: 17, 21, 25, 29, 44, 76, 93

KROPOTKIN, Yelizaveta (Elizabeth), née Korandino (Kropotkin's stepmother): 3-4, 5

KRAVCHINSKY, Sergei: see STEPNIAK

LAND & FREEDOM. See *ZEMLYA I VOLYA*

LANE, Joseph: 56, 61-2, 67

LAVROV, Pyotr: 8, 22, 31, 36, 38, 39, 43-4, 47, 75

LAVROVA, Sofia: 21, 24, 26, 31

LEFRANÇOIS, Gustave: 30, 53

LENIN, Vladimir Ilyich; 36, 99, 101-2

MAKHNO, Nestor: 98

MALATESTA, Errico: 52, 53, 56, 61, 62, 69, 74, 88, 89-92

MALON, Benoît: 30, 31

MANN, Tom: 70, 79

MARX, Karl: 24-6, 27, 29, 37, 39, 60

MARXISTS/MARXISM: 24, 26, 27, 31, 49-50, 60, 83, 89, 95, 96-7

MICHEL, Louise: 30, 53, 56, 69, 82

MIKHAILOV, Mikhail: 16

MORRIS, William: 61, 62, 66-7, 68, 70, 72

MUTUAL AID, 17, 67

MUTUALISM: 24 See PROUDHON

NARODNAYA VOLYA (The People's Will): 33, 56

NARODNIK: 23, 36, 40, 41 See POPULISM

NATANSON, Mark: 40

NECHAEV, Sergei: 35, 36, 40, 41

NIHILISM: 33-36

NIKOLSKOYE, Kropotkin country estate: 2, 3, 5, 23

OBSHCHINA: 37 See PEASANTS

PEASANTS: 1, 2, 10, 11, 37, 38, 41, 43, 47-8, 49, 76, 80, 96, 97, 101
   See SERFS; COMMUNES, PEASANT

PEROVSKAYA, Sophia: 41, 42

PISAREV, Dimitri: 33, 46

POLISH INSURRECTION: 11, 13, 16-17

POPULISM (*NARODNICHESTVO*): 33, 36-39, 40, 43, 47-8, 74

POSITIVISM: 9, 33, 34, 36

PROUDHON, Pierre Joseph: 17, 23, 37, 38, 63, 64

RECLUS, Elisée: 30, 52, 53, 59, 64, 71, 74, 82

REVOLUTION IN RUSSIA: 10, 19-20, 34, 45, 49, 50, 56, 80-81, 93, 97, 99-105

RÉVOLTE, LA (Journal): 55, 56, 67, 71

SAINT PETER & SAINT PAUL FORTRESS: 20, 35, 44, 47-49

SCHWITZGUEBEL, Adhemar: 29, 31

SECOND INTERNATIONAL: 70

SERFS, SERDOM: 1, 2-3, 5, 10-11, 15, 23-4 See PEASANTS

SEYMOUR, Henry: 62-3, 64

SHAPIRO, Alexander: 81, 82, 89, 91, 92, 103, 105

SHAW, George Bernard: 61, 68, 84

SIBERIA, travels in: 13-18

SOCIALISM: 24, 26, 29, 30, 36, 38, 48, 50-53, 61-3, 65, 86, 88, 89-90, 100
    Agrarian socialism: 36, 37, 38, 39, 41, 45
    In England: 60-68, 70
    Libertarian socialism: 29, 67, 68
    Revolutionary socialism: 39, 52, 62, 65, 66, 67, 70, 79
    Social Democrats: 38, 60-63, 70
    State socialism: 26, 27, 28, 30, 38, 50, 61, 87, 97 See MARXISM

SOVIETS (Councils after Russian Revolution): 100, 101, 103

SPENCER, Herbert: 18, 34, 63

STATE, role of: 11, 16, 17, 38, 39, 49, 51, 86, 90, 101, 102

STEPNIAK (Sergei Kraschinsky): 21, 42-3, 48, 49, 60, 61, 64, 72

SYNDICALISM: 79-80, 83

TERRORISM: 33, 35, 41, 45, 56, 70-71, 78, 81, 84

TOLSTOY, Leo: 23, 43, 69, 73, 75-6, 103

TSAR ALEXANDER II: 6, 10-11, 20, 33, 35, 42, 56

TSAR NICHOLAS II: 2, 3, 5, 6, 10, 33 ,84

TSARIST REGIME: 11, 16, 18, 20, 43, 70, 73, 75, 80-81, 84, 93

TURGENEV, Ivan: 33, 39, 75

UTIN, Nicholas: 27-8, 31

WILSON, Charlotte: 63, 64

WORLD WAR ONE: 86-9, 91-2, 96, 100

ZEMLYA I VOLYA (*Land & Freedom*): 35, 40, 42

ZHUKOVSKY, Nicholas: 28, 31, 53